Radioactive Waste Management in International and European Legal Perspective

Radioactive Waste Management in International and European Legal Perspective

Harry H.G. Post and Aurelien Raccah (Eds.)

eleven
international publishing

Published, sold and distributed by Eleven International Publishing
P.O. Box 85576
2508 CG The Hague
The Netherlands
Tel.: +31 70 33 070 33
Fax: +31 70 33 070 30
e-mail: sales@elevenpub.nl
www.elevenpub.com

Sold and distributed in USA and Canada
International Specialized Book Services
920 NE 58th Avenue, Suite 300
Portland, OR 97213-3786, USA
Tel.: 1-800-944-6190 (toll-free)
Fax: +1 503 280-8832
orders@isbs.com
www.isbs.com

Eleven International Publishing is an imprint of Boom uitgevers Den Haag.

ISBN 978-94-6236-694-7
ISBN 978-94-6274-588-9 (E-book)

Printed in The Netherlands

Table of Contents

Abbreviations

AG	Advocate General
AJIL	American Journal of International Law
CEE	Communauté économique européenne
CEZ	The Czech State
CYEL	Canadian Yearbook of Environmental Law
EAEC	European Atomic Energy Community
EC	European Community
ECCS	European Community for Coal and Steal
ECHR	European Convention of Human Rights
ECJ	European Court of Justice
ECLI	European Case Law Identifier
ECR	European Community Reports
ECSC	European Coal and Steel Community
ECtHR	European Court of Human Rights
EEC	European Economic Community
EGC	European General Court
EIA	Environmental Impact Assessment
EIoP	European Integration online Papers
ELD	Environmental Liability Directive
ENSREG	European Nuclear Safety Regulators Group
EP	European Parliament
ERC	European Research Council
ESF	European Social Fund
EU	European Union
EWCA	England and Wales Court of Appeal
EnBW	Energie Baden-Württemberg
Euratom	European Atomic Energy Community
IAEA	International Atomic Energy Agency
ICJ	International Court of Justice
ICRC	International Committee for the Red Cross
IMO	International Maritime Organization
INFCIRC	Information Circulars of the International Atomic Energy Agency (IAEA)
ITLOS	International Tribunal for the Law of the Sea
NATO	North Atlantic Treaty Organization

NGO	Non-governmental organization
OECD	Organization for Economic Co-operation and Development
OJ	Official Journal of the European Union
OSS	Open Systems Specialists
PIC	Prior informed consent
REM	Radiological Environmental Monitoring
SCI	Site of Community Importance
TEC	Treaty Establishing the European Community
TEU	Treaty on the European Union
TFEU	Treaty on the Functioning of the European Union
UK	United Kingdom
UNEP	United Nations Environment Programme
WCED	World Commission on Environment and Development
WEEE	Waste Electrical and Electronic Equipment Directive
WHO	World Health Organization

Preface

According to the *Handbook on Nuclear Law* published by the International Atomic Energy Agency (IAEA) in 2003, nuclear law can be defined as 'the body of *special* legal norms created to regulate the conduct of legal or natural persons engaged in activities related to fissionable materials, ionizing radiation and exposure to natural sources of radiation' (emphasis added).[1] The precise meaning of this speciality has been the subject of some interesting studies by nuclear lawyers, in particular as regards the relationship between nuclear law and environmental law. Although the concept of *lex specialis* is one of the traditional legal tools used to solve conflicts of legal norms, in the sense that *lex specialis derogat legi generali*, most nuclear lawyers who have paid attention to this issue have argued that the relationship between nuclear law and environmental law is not a conflictual one, but rather one of complementarity.

The aim of nuclear law is in fact, as the aforementioned IAEA *Handbook* itself points out, 'to provide a legal framework for conducting activities related to nuclear energy and ionizing radiation in a manner which adequately protects individuals, property and the environment'. It may, therefore, be argued that the objective of nuclear law is, within its special scope of application, similar to that of the largely 'younger' body of law now known as environmental law. On the other hand, there is no doubt that the focus of nuclear law was originally on the protection of people and property, and that the protection of the environment only later became an issue of direct concern, in part due to the influence of the quickly expanding rules of environmental law. Some nuclear lawyers have, therefore, argued that the importance of environmental law for nuclear activities is increasing and may lead to a growing symbiosis with nuclear law.

In this context, this book, which specifically deals with radioactive waste management in international and European legal perspective, should be of great interest to the nuclear lawyer. Although the focus of the book is in fact European law, the issues relating to the relationship between the special rules on radioactive waste management, adopted within the framework of the European Atomic Energy Community (Euratom), and the more general rules on the protection of public health and the environment, adopted within the framework of the European Union, may be seen as largely mirroring similar issues relating to the relationship between nuclear law and environmental law within a State's domestic legal system. Moreover, from the international legal perspective, two chapters in the book explore the relationship between the special Euratom rules on radioactive waste manage-

1 This publication is available on line at <www-pub.iaea.org/MTCD/Publications/PDF/Pub1160_web.pdf>.

ment and, respectively, international environmental law and international human rights law, an area which has so far not attracted much attention by nuclear lawyers.

Andrea Gioia
Vienna, May 2016

List of Contributors

Andrea Gioia	Senior Legal Officer, International Atomic Energy Agency, Vienna, Austria; Professor of International Law, Dipartimento di Giurisprudenza, Università degli Studi di Modena e Reggio Emilia, Modena, Italy
Erik V. Koppe	Assistant Professor, Grotius Center for International Legal Studies, Leiden Law School, The Netherlands
Leonardo Massai	Visiting Professor, Faculté libre de droit, Université Catholique de Lille, France
Mario Odoni	Assistant Professor, Dipartimento di Giurisprudenza, Università degli Studi di Sassari, Sardinia, Italy
Harry H.G. Post	Professor of International Law, ret., University of Exeter, UK; Visiting Professor, Faculté libre de droit, Université Catholique de Lille and Dipartimento di Giurisprudenza, Università degli Studi di Sassari, Sardinia, Italy
Aurelien Raccah	Associate Professor and Vice-Dean for International Development, Faculté libre de droit, Université Catholique de Lille, Lille, France
Stefano Silingardi	Assistant Professor, Dipartimento di Giurisprudenza, Università degli Studi di Modena e Reggio Emilia, Modena, Italy
Miguel Sousa Ferro	Professor at University of Lisbon Law School and at European University, Lisbon, Portugal

The Management of Radioactive Waste

The management of radioactive waste in Europe and elsewhere has been the subject of continuous legal and political challenges, both in court rooms and by environmental groups and NGO's in the field. The transport of such waste, in particular, has often caused considerable public concern and has led to blockades of roads, police interventions and more. One of the purposes of this book is to examine and discuss some important legal aspects of the management of radioactive waste, notably regarding the state of the relevant Law of the European Union.

One of the basic aims of the general policy of the EU regarding waste disposal is 'self-sufficiency', more in particular that the Member States become self-sufficient individually. In order to achieve such self-sufficiency the Member States shall draw up waste management plans and, more in general abide by the relevant European law. Article I of Directive 2008/98, the 'Framework' directive on waste, '(...) lays down measures to protect the environment and human health by preventing and reducing the adverse impacts of the generation and management of waste and by reducing overall impacts of resource use and improving the efficiency of such use'. The Directive addresses waste management in general, but also more specific matters that were previously covered by separate regulation. 'Hazardous waste', for example is now addressed in the Framework Directive. This more comprehensive approach of the Community is a result of the widely ranging Thematic Strategy on the Prevention and Recycling of Waste which was adopted in 2005. In both this Strategy and the Directive, the protection of health and the environment has become the primary purpose prevailing over the trade in waste as a good and a product. Still, tension between those purposes has not vanished as is also illustrated by the rather complicated and very special regime of radioactive waste (and its transport).

Surprisingly perhaps, the management of radioactive waste is not addressed in Directive 2008/98. In respect to this part of EU Waste Law, the Thematic Strategy and the Framework Directive have not led to a change in EU Law. In 1957 Euratom, was created as a separate European Community with special powers regarding the safety of sources of ionizing radiation. Eventually the management of radioactive waste has also been primarily brought under the Euratom regime. However how does this traditional *lex specialis* regarding the management of the 'sources of ionizing radiation' relate to the more general responsibility for public health and the environment as it has gradually overtaken the trading approach to waste and is now expressed in the 2008 Framework Directive and in other EC instruments? At first sight the Euratom regime does not seem to react much to such developments. The Euratom Treaty has hardly been affected at all by the developments from the original Community Law to modern EU Law. The Treaty has not been changed in a substantial way during the policy and legal changes that took place in the period from the 1985 Single European Act, via Maastricht and Amsterdam to the Lisbon Treaty. But in the meantime, it was decided to place the management and shipment of radioactive waste under Euratom instead of under the relevant environmental and public health regime of the EU and TFEU. How does this Euratom regime function in respect to such an important part of the management of waste as radioactive waste is. The 'Euratom choice' brings up interesting legal and policy questions, like how does the modern Euratom legal regime – and therefor EU Law in general – regarding radioactive waste management relate to modern international legal principles that express a 'due care' obligation of States in respect to health and the environment? Such an obligation seems also to underlie (recent) jurisprudence of the European Court of Human Rights, as in the 2004 *Őneryildiz* and the 2012 *Di Sarno* cases. In this book answers to these and other questions will be sought in a discussion of the recent Euratom Directives 2011/70 on the safe management of spent fuel and radioactive waste (it is in force since 23 August 2013) and 2006/117 on the Supervision and Control of Shipments of Radioactive Waste and Spent Fuel. This will bring up also more practical legal issues like: Do Member States have the right to refuse the transit of radioactive waste transports under the latter Directive, or what conditions can they impose to such transports (Article 9.3(a))? The production and the transport of radioactive waste have led to a variety of political and legal problems and questions also in respect to the world *outside* the European Union. The rather extensive regime of the 1989 Basel Convention on the Control of Trans-boundary movements of Hazardous Wastes and their Disposal, e.g., does not cover the transport and disposal of radioactive waste.

In Chapter 1 *Harry Post* discusses the considerable intricacies of the law regulating the management of radioactive waste in general and therewith most of the issues that are addressed in the subsequent chapters of the book. Following the introduction of the basics of the EU Law regarding the management of waste, in general, and of radioactive waste, in particular, he continues by pointing to the remarkable *ius specialis* character of Euratom Law. It puts centrefold the potential hierarchical nature of the relationship between Euratom Law and General EU Law, the subject of the next chapter by Miguel Sousa Ferro. It leaves a central question: to what extent – if at all – does EU Treaty Law in its modern form apply to the nuclear energy sector? Next, he addresses some of the changes in the Euratom regime notably brought by the new 'Framework' Directive 2011/70/Euratom on Safe Management of Spent Fuel and Radioactive Waste, and, finally, he outlines the regime regarding the transport of radioactive waste. He ends his chapter by posing some critical questions. Post expresses some confidence in the direction of reform of the EU/Euratom regime of radioactive waste management that is under way (i.e. more information and public participation), but he fears that a long way is still to be travelled before Euratom has really effectively been changed in this way. He doubts that in view of the potential dangers involved, relying largely on 'self-assessment' as the new Euratom Framework Directive does, will be adequate to ensure a high standard of care for the management of radioactive waste (and of spent fuel).

In Chapter 2 *Miguel Sousa Ferro*, analyses the Euratom regime of the management of radioactive waste in detail. Although the Euratom Treaty itself does not explicitly foresee the adoption of binding rules on the management of radioactive waste, the functional and wide interpretation by the European Court of Justice provided enough justification for the adoption of two Directives relating to nuclear safety and radioactive waste management and of other instruments relevant for the management of radioactive waste. Sousa Ferro provides a deeply digging analysis of the *ius specialis* character of the Euratom. He argues that the real issues are the identification of the precise consequences of this relationship of *lex specialis – lex generalis*, and how the two are to be articulated. His detailed analysis focuses on the interpretation of Euratom provisions, the application of provisions of TEU/TFEU primary or secondary legislation to radioactive waste and the choice of the correct legal basis for legislation on radioactive waste. In respect to the matter of the interpretation of the Euratom regime, he emphasizes the shift in the jurisprudence from a broad, functional approach focussing on the protection of public health to a more restrictive approach, e.g., excluding the application to military activities. As the major reason behind this change Sousa Ferro sees the failed Convention on the Future of Europe which on the one hand, eventually led to the 2009 Lisbon Treaty but, on the other hand, did not lead to substantial changes in the Euratom Treaty, and thus to a more comprehensive EU. He proceeds then by analysing the (potential) consequences of this failure to create a single EU legal order, including the uncertainties about the application of TEU/TFEU legislation to the management of radioactive waste. However, in that latter area, he cannot detect, nor immediately foresee, great problems: in principle TEU/TFEU primary and secondary legislation apply to radioactive waste, unless the issues at stake are regulated by the Euratom Treaty in a way which is incompatible. Such (potential) incompatibility is hard to foresee. The choice of a correct legal basis for legislation on radioactive waste, in Sousa Ferro's view, causes more problems, in particular when the issue of military radioactive waste and its transport is at stake. Recent jurisprudence after all seems to exclude military activities from the scope of the Euratom regime. However as the primary goal of legislation on radioactive waste remains protection from the dangers of ionizing radiation, for the management of such military waste the Court's own jurisprudence would nevertheless only allow legislation with a legal base in the Euratom Treaty. Therefore, the only solution he envisages is the creation of a new instrument. Its legal basis however, is likely to be challenged whether it is in the TFEU or in the Euratom Treaty, or in both treaties. The latter solution is not really feasible as the procedures under both Treaties are different, notably due to the merely consultative role of the European Parliament under Euratom, and never changed since its creation. He also does not believe that a TFEU legal basis would be acceptable in view of the Court's position that protection from ionizing radiation is exclusively a Euratom matter. Moreover, he already questions the legitimacy of TFEU legislation affecting radioactive waste transport, like the one on Environmental Impact Assessment.

In Chapter 3 *Stefano Silingardi* proposes a further analysis of the Euratom jurisprudence of the European Court of Justice. Beforehand, he expresses surprise about the limited size of the Euratom jurisprudence, but eventually does conclude that 'it is unquestionable that the ECJ has played a decisive role in shaping EU Nuclear Law'. In the first place, the Court has basically removed legal obstacles the European Commission seemed to have perceived as a barrier to the adoption of legislation on the safety of nuclear installations and on radioactive waste. Secondly, the Court also set a limit to the powers of the Commission in respect to nuclear activities: it held that Euratom law, generally, does not apply to military installations and activities. The net result, Silingardi argues, is at this point rather unsatisfactory. The Court ruled that the Member States should undertake legislative action based on the TFEU, instead of Euratom. It seems unlikely that they will take up this challenge leaving the EU citizens basically unprotected to the dangers of military nuclear activities including those that produce radioactive waste. In another part of its jurisprudence, the ECJ seems to be assuming a rather central role in respect to subjects, like nuclear safety, where the interests of protecting the public can clash with national security interests. The EU executive is reluctant to enter such areas and, in respect to the Euratom regime the European Parliament simply has no power. The Court seems to take on the role of finding an appropriate balance here. It is to be seen what the result will be.

Leonardo Massai analyses the legal regime under Euratom and other EU Law in respect to the transport of radioactive waste. In this Chapter 4 Massai, firstly, draws attention to the sheer volume and size of the (low level) radioactive waste that is routinely transported globally and in Europe – his focus – and the exclusion of the transport of radioactive waste from the legislation of the transport of other waste. Shipments of radioactive waste (and spent nuclear fuel) between Member States are primarily regulated on the basis of Directive 2006/117/Euratom which most important provisions and procedures are examined. The Directive establishes a system of prior authorization for such shipments in Europe, shipment to countries which do not have the resources to safely manage is prohibited, and the regime for transport *intra* the EU differs from that for transport outside the EU. He, furthermore, discusses the additional binding conditions for the transport of radioactive waste of the 2011/70/Euratom 'Framework' Directive on Radioactive Waste. Massai concludes his chapter by discussing the proposal for a Community System for Registration of Carriers of Radioactive Materials and the 2013 (first) Commission Report on the implementation by the Member States of Directive 2006/117/Euratom. The implementation of the Directive has not given rise to major problems, although the lack of sufficient harmonisation of the procedures of Member States was reason for some concern.

Chapter 5 by *Mario Odoni* discusses radioactive waste management in the light of human rights obligations. The author refers in some detail to the proceedings in the case of the Quirrra Test/Training Range, a huge military testing area, currently before the Court of Lanusei in Sardinia. Serious environmental and health concerns have been raised linked to the prolonged military use of the area. He reviews the range of questions on the Quirra being asked by Members of the European Parliament and what, in his view, were the rather unsatisfactory answers of the European Commission. The Commission bases its answers repeatedly on relevant ECJ case law which the author then briefly proceeds to discuss. Odoni is critical of the 2005 and 2006 judgments in which the Court concludes that the Euratom Treaty is not applicable to military activities, in particular, he questions the view that even *the possible effects* of uses of nuclear energy for military purposes, namely their negative impact on human health and environment, are by definition *outside* the scope of the Treaty. He bases his view on a detailed analysis of the texts at issue, including the Euratom Treaty as well as relevant provisions of the TFEU and TEU and argues that in respect to this subject Member States cannot exercise their competence in such a way as to jeopardize the general Euratom/Union objective to protect the environment and public health. Odoni ends his chapter by applying his findings to the *Quirra* case in which he includes an analysis of the European Convention of Human Rights. He concludes that no rule under EU/Euratom Law could be interpreted as derogating the *erga omnes* customary duty of Member States to prevent, reduce and control harm even to their own domestic environment. This duty he, finally, submits includes the obligation to inform the public, without delay, on possible risks arising from exposure to a contaminated environment and that seems to be the situation of the territory around the Quirra military test/training range.

In the last chapter, Chapter 6, *Erik Koppe* discusses to what extent Euratom is bound by general rules of International Environmental Law. He focuses on the emergence of a customary duty of care for the environment also applicable to Euratom. Firstly, he analyses the Euratom Treaty and argues that its founders were clearly intended to endow it with international legal personality and provided rights and powers accordingly. Euratom is therefore bound by rules of Customary International Law, including the rules of Customary International Environmental Law. Koppe proceeds by an overview of the development and the current state of these rules and then carefully analyses in particular the General Principles on which the field is founded, like those of prevention, of precaution, and of good neighborliness. How are such principles identified? is a question he answers, firstly, in respect to the Prohibition to cause (significant) trans-boundary pollution and, secondly, to the related Duty of care for the environment. He believes that these two Principles can be amalgamated into one single norm: a general and customary duty of care towards other states, more specifically a general duty of care for the environment in other states and for the environment in areas beyond national jurisdiction. The latter is Koppe's central concern in his chapter. He provides arguments supporting the existence of such a general duty of care for the environment not only from a careful analysis of international jurisprudence but also argues that there is sufficient ground for its identification in the *opinio juris*. Finally, he draws attention to the fact that it is generally accepted that states have an obligation under Customary International Law to protect the intrinsic value of the environment in times of armed conflict. *A fortiori* such duty of care would apply in times of peace.

Two workshops have served as a major source of inspiration for the book that lies here before you. 'The Management of Radioactive Waste and the Duty to Care for Health and Environment' was the title of the workshop that took place in Sassari, Sardinia, on Friday 17 May 2013. This workshop was made possible by the Università degli studi di Sassari and funded by the European Commission through the Centro di documentazione europea per l'Anno europea 2013. The University of Sassari organised the meeting in a very professional and friendly way. The presence of Professor Emeritus Attilio Mastino, Magnifico Rettore of the University, as well as his inspired speech were a particular honour. We are most grateful to dottoressa Magda Sanna of the Centro di documentazione europea, to Professor Silvia Sanna of the Dipartimento di Scienze politiche, Scienze della comunicazione e ingegneria dell'informazione, and to Professor Paolo Fois, who competently chaired the meeting, and Dr Mario Odoni of the Dipartimento di Giurisprudenza. Without their efforts there would have been no workshop.

On 4 April 2014 an International Workshop/*Journée d'étude internationale* took place in Lille, France, entitled 'Radioactive waste: The Management of Radioactive Waste and its Transport in European and International Legal Perspective'. The workshop was hosted by the *Faculté libre de droit* of the *Université catholique de Lille* and supported and funded by the Faculty and the *Fondation Cardinal Poupard*. Particular thanks go to the Dean of the *Faculté libre de droit*, Dr Ioannis Panoussis and the members of the International and European Law Department (IELS) and the *Global Legal Network* (GLN) for their precious help. Ms. Zuzana Babikova deserves special thanks for the organization of the workshop.

Both the workshops in Sassari and the one in Lille were considered very successful meetings as a result of the highly competent group of experts that took part and the participation, in both cases, of an enthusiastic and interested audience. The contributions demonstrated convincingly how important but also how difficult proper and acceptable management of radioactive waste is. Gratitude is due to each of its contributors and also to those who have generously supported the organization of the workshop and the publication of the book. We would further like to express our gratitude to the *Fondation Cardinal Poupard* which has generously supported both the workshop in Lille and the publication of this book. We also like to thank the staff of Eleven International Publishing for their competent help (and patience) during the publication process of this book, and Ms. Dina Sušanj for her thorough language revision of the manuscript.

Harry H.G. Post and *Aurelien Raccah*
Lille-Sassari, March 2016

1 Radioactive Waste Management under Euratom and Other EU Law: An Introduction

Harry H.G. Post

For many years, the management of radioactive waste has been subject of continuous concerns and legal as well as political challenges, in particular by environmental groups and NGO's in the field, and to a lesser extent in courts of law, like currently on Sardinia in the Court of Lanusei.[1] Often these challenges concerned the management (and the transport) of spent nuclear fuel. 'What to do with spent nuclear fuel?' is, still in 2014, as it has already been for decades, one of the big technical questions of the management of nuclear energy.

In other cases, challenges regarded other forms of radio-active waste, like emissions (gaseous, waste water) from nuclear plants, including accidental emissions due to deficiencies. In still other cases, the cause of concern, like before the Court of Lanusei is the military use of munitions and other devices which (may) include radio-active materials. In all such cases the basic worries regard dangers for the health (of the general public, of workers) and the protection of the environment.

It is my purpose to introduce some legal aspects of the management including the transport of radioactive waste in Europe. A further and much more profound and interesting analysis of many of the aspects of the legal regime of radioactive waste will be provided by the contributions that will follow in this book.

Here, I will first address some basic characteristics of the modern regime of European waste management in general and of radioactive waste management in particular (Section 1.1). Next, I will introduce the somewhat remarkable *ius specialis* character of radioactive waste management in EU Law, focussing on the rules on nuclear safety and environmental liability (Section 1.2). In Section 1.3, I will address relatively recent and very recent changes in the Euratom regime, as will be done much more profoundly, further on in this book. These developments are brought by new legal instruments, notably by the new 'Framework' Directive 2011/70/Euratom on safe management of spent fuel and radioactive waste.[2]

1 The case at the Court in Lanusei concerns also all kinds of other waste issues. See Chapter 5 by Mario Odoni in this volume.

2 Council Directive 2011/70/Euratom of 19 July 2011 Establishing a Community framework for the responsible and safe management of spent fuel and radioactive waste, in OJ L 199/48 of 02.08.2011.

Section 1.4 will introduce the rather complicated legal regime regarding the transport of radio-active waste, also to be explained more profoundly below. The last section contains some reflective conclusions.

In the process I hope to show that the law regarding radio-active waste is not only very important but also rather curious, and probably just as 'strangely engaging' as European Waste Law in general.[3] This is perhaps in part so because to a large extent the dynamics of modern International Environmental Law seem to have bypassed European Law on the management of radioactive waste. One may perhaps even say that it has been left in curious and perhaps not even 'splendid isolation'. However, I will not only argue that I consider it time for a revision of in particular, the Euratom regime of radioactive waste management, but I will also submit that there is already some initial support available for the thesis that 'the prince has entered the palace and is perhaps prepared to give this not so very beautiful 'Sleeping Beauty' a kiss.... '

1.1 Some Basics

1.1.1 Management of Waste

One of the principal aims of the policy of the EU regarding waste is 'self-sufficiency'. More in particular this means that the Member States are supposed to attain individual self-sufficiency. In order to achieve such self-sufficiency the Member States shall draw up waste management plans and, of course, more in general abide by the relevant European Law. Article I of Directive 2008/98, the 'Framework' Directive on Waste,

> (...) lays down measures to protect the environment and human health by preventing and reducing the adverse impacts of the generation and management of waste and by reducing overall impacts of resource use and improving the efficiency of such use.

The Framework Directive addresses waste management in general, but also more specific matters that were previously covered by separate instruments. 'Hazardous' waste, for example, is now covered explicitly and in some detail in this Directive. This more comprehensive approach of the EU is a result of the 'Thematic Strategy on the Prevention and Recycling of Waste' which was adopted in 2005. In both the Strategy and the Framework

3 Martin Lee, cited in Jan Jans & Hans Vedder, European Environmental Law, 4th edn, Groningen, Europa Law Publishing, 2012, p. 213.

Directive, the protection of health and the environment have become the primary purpose prevailing over the trade in waste as a good and a product.

The regime regarding the management of radioactive waste is not addressed in the Framework Directive although some of it is certainly hazardous. It actually has, as mentioned above, its 'own' Framework Directive, Directive 2011/70/Euratom. This instrument establishes 'a Community Framework for the responsible and safe management of spent fuel and radioactive waste'. It was adopted on 19 July 2011 and was supposed to be transposed by the Member States ultimately on 23 August 2013. Before turning to some remarks about this basic framework for regulation, let us, first, look at some of the general aspects of European Radioactive Waste Law.

1.1.2 *Waste and Radioactive Waste*

For the purposes of the 2008 Waste Framework Directive (2008/98/EC) 'waste' means: '(...) any substance or object which the holder discards or intends or is required to discard; (...)'.[4] This short phrase has led to endless questions and arguments in numerous cases before the European Court of Justice to the extent that an experienced judge reported his honest 'amazement' about the judicial effort put into defining the concept of waste.[5] In its 1997 judgment in the *Inter Environnement* case, the Court – although negatively – chose a broad approach deciding '(...) that the concept of waste does not in principle exclude any kind of residue, industrial by-product or other substance arising from production processes'.[6] The wise ladies and gentlemen from Luxembourg undoubtedly did not believe that they had solved the puzzle of providing a definition able to keep up with continuous technological innovations, but this formula at least could cover a lot.

Radioactive waste is explicitly excluded from the general 2008 Waste Framework Directive as it was from its predecessors. Its 'own' instruments define radioactive waste as

> (...) radioactive material in gaseous, liquid or solid form for which no further use is foreseen or considered by the Member State or by a legal or natural person

4 Directive 2008/98/EC of the European Parliament and of the Council of 19 November 2008 on Waste and Repealing Certain Directives, OJ 2008 L 312/3, Art. 3.1. The preceding Waste Framework Directive (2006/12/EC) in Art. 1.1.a had the same definition but with a reference to categories of waste set out in its Annex I. Annex I ended, in Q 16, with a 'catch-all' category: 'Any materials, substances or products which are not contained in the abovementioned categories.' This effectively declared that all substances not enumerated should also be declared waste.

5 J. Tieman, *Naar een nuttige toepassing van het begrip afvalstof: over de betekenis en toepassing van kernbegrippen van internationaal, Europees en Nederlands afvalstoffenrecht* [Towards a Useful Application of the Concept of Waste: On the Meaning and Application of Core Concepts of International, European and Dutch Waste Law] [in Dutch], Kluwer, Deventer, The Netherlands, 2003, pp. 5 et seq. Tieman's book by itself provides quite an illustration of Judge Kapteyn's finding.

6 Case C-129/96, *Inter-Environnement Wallonie ASBL v. Waals Gewest* [1997] ECR I-7411.

> whose decision is accepted by the Member State, and which is regulated as radioactive waste by a competent regulatory authority under the legislative and regulatory framework of the Member State;

Radioactive waste in this sense can include 'spent fuel' as Article 3(11) says that:

> (...); spent fuel may either be considered as a usable resource that can be reprocessed or be destined for disposal if regarded as radioactive waste;[7]

What strikes in this definition in comparison to the simple general definition of waste of the (2008) Waste Framework Directive is the inclusion of the decision on 'no further use' within the definition. In that sense, the question can even be asked whether the substantive part of the definition of radioactive waste of Article 3(7) does not lose much of its meaning. The public authorities, after all, keep a considerable freedom this way. They can even choose to accept the decision of 'a legal or natural person' as long as a competent regulative authority controls the radioactive waste so determined. Different than in the case of the general waste definition, in its limited number of cases regarding radioactive waste the ECJ has not further clarified or otherwise elaborated upon the scope of the definition.

Why this separation, this double approach to waste in EU Law? Radioactive waste certainly has a rather specific character as a substance. There are specific dangers and there is a specific security side to at least a part of it involved. Still one might immediately oppose: there is other 'hazardous' waste which is also rather dangerous. The major explanation, I believe, for the legally rather remarkable nature of radioactive waste management in Europe, is the curious background and legal history of its creation and birth pains, as well as the regulation of nuclear affairs chosen and enacted in the first decades of the European Communities.

1.2 Euratom: A Special Community

In 1957, Euratom was founded in the Treaties of Rome that brought us also the European Economic Community. Together with the 1952 European Community for Coal and Steel, ECCS, Euratom was one of the three original communities agreed upon and created to 'carry' European economic integration. However whereas the ECCS and the EEC have disappeared and amalgamated into the European Union, Euratom continues to exist as a special Community within the EU.

7 Council Directive 2011/70/Euratom of 19 July 2011, above n. 2, Art. 3(7). Art. 5(1) and (2) of Council Directive 2006/117/Euratom of 20 November 2006 on the supervision and control of shipments of radioactive waste and spent fuel are similar (in OJ L 337/21 of 05.12.2006).

Just as the Treaty on the European Union and the Treaty on the Functioning of the European Union are binding Primary Law for the contracting parties, so is the 1957 Treaty establishing the European Energy Community (Euratom).

Article 305 EC (before 232) stated the relationship with the Euratom Treaty as follows:

> The provisions of this Treaty shall not derogate from those of the Treaty establishing the European Atomic Energy Community.

This text gives expression to a classic rule of Treaty hierarchy: That of the primacy of the *lex specialis* (Euratom Treaty) over the *lex generalis* (the EU and TFEU Treaty). It would probably be incorrect to argue only based on Article 305 EC that there is or was a hierarchical relationship between the Treaties in the sense that the Euratom Treaty would prevail over the EU Treaties in respect to the subject matter covered in the Euratom Treaty. It is most likely that the two Communities and Treaties are more or less autonomous.[8] The fact that Article 305 EC has been repealed by the Lisbon Treaty strengthens that idea. What is left now is Article 106a(3) Euratom, added to the Euratom Treaty by way of the second Protocol to the Lisbon Treaty. It says:

> (3) The provisions of the Treaty on European Union and of the Treaty on the Functioning of the European Union shall not derogate from the provisions of this Treaty.

In all this, we may also remember that in 1996 the ECJ ruled that where a sectorial Treaty (in this case the ECSC Treaty) deals exhaustively with a topic there is no place for application of the EC Treaty.[9]

Perhaps the most interesting question this co-existence raises, however, is to what extent – if at all – EU Treaty Law in its modern form applies to the nuclear energy sector? This question is even more interesting because the Euratom Treaty has not substantively changed during all the major developments leading to the modern EU.[10] It more or less slept through all the European integration developments and revolutions. It went unscathed through the Single European Act of 1985, the 1992 Treaty of Maastricht, the 1997 Treaty of Amsterdam, the 2000 Treaty of Nice and even in fact through the complicated delivery

8 Thomas F. Cusack, 'A Tale of Two Treaties: An Assessment of the Euratom Treaty', (2003) *Common Market Law Review*, pp. 117-141. Cusack uses autonomous and even 'exclusive'. See pp. 127 and 135.

9 Case C-18/94, *Hopkins et al. v. National Power plc &Power Gen plc*, [1996] ECR I-2281 at 2318, para. 22. Or, as the Commission said in its Decision in the KLE Case: '(...) the European Atomic Energy Community and the European Economic Community were established, from a legal, organizational and institutional viewpoint, as two mutually independent Communities and the legal acts of one Community are not subject to the acts of the other'. (Commission Decision of 21 February 1994, 94/285/Euratom, OJ 1994, L 122/30).

10 Notably the core Chapters i-vi of Title II of the Euratom Treaty have not changed at all since 1958.

of the 2007 Treaty of Lisbon although that led to some institutional changes expressed in the Euratom Treaty. This results in the remarkable situation that the co-decision-making procedure in Euratom has remained in place while decision-making in the EU is now completely different. Hence, in the process of enacting secondary Euratom legislation the powers of the European Parliament are very small as compared to in decision-making regarding other EU legislation. Only recently has this curious difference – for the first time – led to an application by Parliament against the Council.[11] In its action, the European Parliament seeks the annulment of a new Euratom Directive 2013/51/Euratom with regard to radioactive substances in water intended for human consumption, *inter alia,* by challenging the choice of the Euratom Treaty as the legal basis for the Directive, instead of Article 192 TFEU. Parliament argues that the measures covered by the Directive fall within the responsibility of the EU for the protection of the environment.

Apart from these 'internal' developments, Euratom seemed also not to have been touched by the emergence and further changes in General International (Environmental) Law after the 1992 Rio Declaration, such as notably the 1998 Aarhus Convention on *Access to Information*, Public Participation in Decision-Making and Access to Justice in Environmental Matters, or the emergence of modern international legal principles that express a 'due care' obligation of States in respect of health and environment. Whereas some of the international developments, in particular perhaps the implementation of the Aarhus Convention in the EU by way of several Directives[12] and Regulations, including those resulting from the fact that the EU has become a Party to the Convention in her own right,[13] have had a profound impact on EU Environmental Law, until recently no such impact on Euratom could be noticed.

Euratom Law was more affected, although perhaps somewhat selectively at times, by developments in International Nuclear Law such as the adoption of the 1994 Nuclear Safety Convention and the conclusion of the 1997 IAEA Joint Convention on the Safety of Spent Fuel management and on the Safety of Radioactive Waste Management.[14] In 2005, the EU

11 Action brought on 30 January 2014, *European Parliament v. Council of the European Union* (Case C-48/14); see, the Chapter 2 by Miguel Sousa Ferro in this volume.

12 Directive 2003/4/EC of the European Parliament and of the Council of 28 January 2003 on public access to environmental information (repealing Council Directive 90/313/EEC), and Directive 2003/35/EC of the European Parliament and of the Council of 26 May 2003 providing for public participation. So far, the third 'pillar' of the Convention, Access to Justice, has not found expression in secondary legislation, although the Commission has tabled its proposal already in 2003 (COM(2003)624 final).

13 The EC is a Party to the Convention since May 2005. Regulation (EC) No. 1367/2006 of the European Parliament and of the Council on the application of the provisions of the Aarhus Convention on Access to Information, Public Participation in Decision-Making and Access to Justice in Environmental Matters to Community institutions and bodies (OJ L 264, 25.09.2006, p. 13) entered into force on 28 September 2006 and became applicable on 17 July 2007. Art. 3 of this 'Aarhus Regulation' declares applicable the already existing Regulation 1049/2001 on access to information held by EU institutions.

14 See Miguel Sousa Ferro's critical comments on the way the Nuclear Safety Convention has been 'implemented' by the EU: 'Directive 2009/71/Euratom: The Losing Battle against Discrimination and Protection of

became a Party to the Joint Convention. What kind of legal regime regarding the management of radioactive waste is then the result of this development?

Euratom has very extensive powers regarding the supply and property of basic raw nuclear materials and in respect of the free movement of goods and persons in the nuclear sector. The Community manages the nuclear common market in Europe including supporting research and investment. Furthermore, it has explicit competence regarding the safety of sources of ionizing radiation, and it has explicit powers in respect of Nuclear Safeguards (Chapter 7) as well as the protection of the health and safety of workers and of the general public against the dangers of ionizing radiation (Chapter 3 of the Treaty).[15]

The Treaty does not give such powers regarding nuclear safety and neither does it provide competence for the management of spent fuel and radioactive waste. However, Euratom is supposed to be generally competent in the field of nuclear safety, including by the European Court of Justice.[16] It has enacted considerable secondary legislation also on radioactive waste management for which the authority of the Court is also referred to. The new Directive 2011/70/Euratom, just mentioned, establishes even 'a Community Framework for the responsible and safe management of spent fuel and radioactive waste'.[17] In the Preamble, the new Directive refers explicitly to the authority of the Court when it states that the provisions of Chapter 3 on Health and Safety of the Euratom Treaty 'form a coherent whole conferring upon the Commission powers of some considerable scope in order to protect the population and the environment against the risks of nuclear contamination'.[18]

In 1992, in its Judgment in *Commission v. Council*, the Court also stated, different than the Council, that as far as EU competences were concerned:

> it is not appropriate, in order to define the EC's [now Union's] competences, to draw an artificial distinction between the protection of the health of the general public and the safety of sources of ionising radiation.[19]

Sovereignty', 2009 *Journal of Nuclear Law*, No. 4, pp. 295-312. The IAEA Joint Convention entered into force on 18 June 2001. Since 2005, Euratom has been a Party to the Convention (see Decision 2005/84/Euratom, OJ 2005, L30/10).

15 Euratom is to create a 'nuclear common market' and thereto has explicit competence in respect of the supply of relevant basic raw materials (Ch. 6) and the free movement of goods and persons (Ch. 9). Furthermore, investment in the sector is encouraged (Chs. 4 and 5) and so are research and dissemination of information (Chs. 1 and 2).

16 See C-29/99 *Commission v. Council*, ECR (2002) I-11221.

17 Directive 2011/70/Eurotom, above n. 2, does not apply to radioactive mining waste, which is covered by the 2006/21/EC Directive on the Management of Waste from Extractive Industries, OJ 2006 L 102/15.

18 Above n. 2, which concurs with the Judgment in C-187/87, *Saarland and Others, v. Min. de l'Industrie* (Catténom), ECR (1988) p. 5013), OJ 2011 L 199/48.

19 C-29/99, above n. 16, para. 82. The Council stated that the Euratom Treaty did not provide the Union with competence to regulate the opening and operation of nuclear facilities as that competence remained with the Member States.

Hence, we must assume that the responsibility for nuclear safety has to be taken broadly, implying responsibility for the safety of spent fuel as well as for the safety of radioactive waste.[20]

Under the Euratom Treaty, the EU has developed a special regime for the management of radioactive waste. As said above, neither the 2006 Waste Framework Directive nor the old Hazardous Waste Directive applied to radioactive waste.[21] That exclusion has not changed in the 2008 Waste Framework Directive. Where precisely the boundaries lie between the competence of the EU for the protection of public health and the environment and Euratom's powers regarding the safety of sources of ionising radiation is therefor far from crystal clear and an important subject of further research. The classic question of the correct legal basis of appropriate secondary legislation may appear here in full force.[22]

Since 1958, risk of radioactive waste pollution from the territory of another Member State and plans for the disposal of radioactive waste are governed by Article 37 Euratom. This Article provides the Commission with the opportunity to deliver an opinion on the possibility of trans-boundary radioactive contamination.[23] Under the Treaty, the Commission has powers to verify the operation and efficiency of nuclear facilities, including those for storing radioactive waste in the Member States. It has not made extensive use of these powers at all. However, this unsatisfactory state of affairs regarding such an important and dangerous matter seems to have reached an end finally on 23 August 2013, the date when the Euratom Directive 2011/71 for the safe management of spent fuel and radioactive waste had to be transposed by the Member States.[24]

20 Also in view of the reference in the Preamble of Directive 2011/ 70 to Council Directive 2009/71/Euratom on the nuclear safety of nuclear installations.

21 Similarly Directive 2004/35/EC on environmental liability with regard to the prevention and remedying of environmental damage (Preamble, para. 10) does not apply to the management of radioactive waste.

22 As mentioned above (n. 11), a case where the issue of the correct legal basis in this regard was explicitly put forward was on 30 January 2014 brought by the European Parliament against the Council (Case C-48/14). See also Case C-376/90, where the ECJ rejected the Commission's argument that Member States were not allowed to enact more stringent protective measures than those of the Euratom regime: *Commission v. Belgium* [1992] ECR I-6153. Moreover, it is also to be seen whether or not there will be tension between the internal market competence and environmental concerns in respect of the EU energy policy as laid down in Art. 194(2) TFEU, affecting the Euratom regime (see Jans &Vedder, above n. 3, p. 86).

23 Art. 37 Euratom was subject to the Court's scrutiny in the *Catténom* case, Case 187/87, 1988 ECR, p. 5013, above n. 18.

24 See Cusack on an attempt by the Commission in the early 1980s to change part of the Euratom regime, in particular Ch. 6 of the Treaty. The Commission considered '(...) that the nuclear fuel-cycle industry had reached a considerable degree of maturity and that the time had come to make it subject, to some extent, to the rigours of EC competition law'. The revision was proposed but not adopted (2003 *Common Market Law Review*, above n. 8, p. 137).

1.3 The New Framework Directive 2011/71/Euratom for the Safe Management of Spent Fuel and Radioactive Waste

Directive 2011/70/Euratom applies the Euratom basic safety standards for the protection against radiation of workers and the general public to the management of spent fuel and radioactive waste.[25] The Directive obliges the Member States to adopt and maintain national policies on spent fuel and radioactive waste management (Article 4), including a national programme for which conditions and criteria are set (Articles 11 and 12). A national regulatory framework for spent fuel and radioactive waste management (Article 5), and a competent national authority for the waste substances concerned (Article 6) have to be established. The new Directive covers all stages of spent fuel management and all radioactive waste management ('from generation to disposal'), but there is an important exception: It only applies to waste (and spent fuel) resulting from civilian activities (Article 2.1). In this respect, it follows the ECJ in its 2005 Judgment in C-61/03 (*Com. v. UK*). Adequate financing for the often very costly long term management of radioactive waste is to be ensured (Article 9), which may involve the nuclear industry. Article 9, entitled 'Financial Resources' says

> (...) that adequate financial resources be available when needed for the implementation of national programmes referred to in Article 11, especially for the management of spent fuel and radioactive waste, taking due account of the responsibility of spent fuel and radioactive waste generators.

Article 10 stipulates a special transparency regime for the sector. Information shall be made available to workers and to the general public which also shall be given the necessary opportunities to participate effectively in the decision-making process regarding spent fuel and radioactive waste. The limits to the provision of information and the participation in decision-making processes are set by national legislation (other interests, like security and commercial competition, should not be jeopardised) and international obligations. Here we see some reflection of the Aarhus Convention and the secondary legislation in that regard that the EU has enacted so far.[26]

Every three years Member States shall report on the implementation of the Directive. Based on these reports the Commission is supposed to submit its own report to Parliament (Article 14(1) and (2)). To complete the stipulations on 'Reporting', Article 14(3) concludes with a remarkably weak enforcement provision:

---

25 With the notable exception of radioactive waste from the extractive industries (covered by its own Directive 2006/21/EC, above n. 17).

26 See, above n. 12 and 13, as well as the text.

> Member States shall periodically, and at least every 10 years, arrange for self-assessments of their national framework, competent regulatory authority, national programme and its implementation, and invite international peer review (...)

They shall report the outcomes of any of such peer reviews to the Commission and the other Member States. Unless there is a conflict with security and proprietary information, these reports shall also be made publicly available.

In view of the potential dangers involved in the management of spent fuel and radioactive waste (notably the still unsolved matter of how to cope with this kind of waste) as well as the huge spending at stake including public funds, this clause may be called rather astonishing.[27]

1.4 TRANSPORT OF RADIOACTIVE WASTE

1.4.1 *Transport of Hazardous Waste*

Regulation (EC) 1013/2006 prohibits the export of hazardous waste to non-OECD countries. In that sense the Regulation implements the 1989 Basel Convention (Decision II/12). If such waste is for disposal only, the Member States may prohibit the import of all 'Regulation' hazardous waste (conform Article 11(1)(e), of the Basel Convention). However, if the import is for recovery purposes there is controversy over the Member States' powers to restrict such imports.

This 2006 Regulation distinguishes 'green' (Annex III) and 'amber' (Annex IV) waste.[28] The 'red' list of the preceding 1993 Regulation has been replaced by a list of specific waste (Annex V) for which export is prohibited, including export of mixtures of hazardous and non-hazardous waste. To non-OECD countries such exports are prohibited whether for disposal or recovery. The composition of these lists in comparison to those of the Basel Convention and other relevant international instruments is a topic for some further research. The distinction between intra-EU transport and transport outside the EU of the 1993 Regulation is maintained in the new instrument.

27 The self-assessments by the Member States concern '(...) their national framework, competent regulatory authority, national programme and its implementation, (...)' the international peer review covers their national framework, competent regulatory authority and/or national programme with the aim of ensuring that high safety standards are achieved in the safe management of spent fuel and radioactive waste'. For the peer review, other than in the case of the self-assessments, there is a choice what is to be covered: either all three subjects of the self-assessment or two of them. The 'implementation' of the national programme is mentioned for self-assessment but not for peer review (Art. 14(3)).

28 For green list waste a general information requirement is stipulated, for amber waste a PIC procedure.

However, the Basel Convention does not apply to shipments of spent fuel or radioactive waste to a third country and neither does Regulation 1013/2006 implementing the Convention.[29] For such a transport, the substantive provisions of the applicable IAEA and IMO Conventions were brought into special EU legislation by way of Directive 92/3/Euratom,[30] replaced by Directive 2006/117/Euratom.[31] In particular, this was the case for the 1997 IAEA Joint Convention on the Safety of Spent Fuel Management and on the Safety of Radioactive Waste Management.[32]

1.4.2 Transport of Radioactive Waste

Radioactive waste is not mentioned in the Annexes of the 2006 Hazardous Waste Regulation. It is regulated separately and so is the shipment of radioactive waste, primarily in Directive 2006/117/Euratom (transposed since 25/12/2008).[33] This 2006 Directive has been modelled on the arrangement for hazardous waste under the Basel Convention just mentioned (e.g., each transport requires notification and permission), and decisions taken in the context of the OECD (the NEA), the IAEA and the IMO. The 2006 Directive maintains the distinction between intra-EU transport and transport outside the EU (as in the Regulations on the Transport of Hazardous Waste). Other than under the old Directive 92/3 Euratom, shipment of spent nuclear fuel for reprocessing is now also included in the 2006 Directive (under Article 1(2)). Member States of transit may refuse consent if they consider the shipment contrary to their international, European or national obligations (Article 9(3)(a)). Member States of destination may also invoke legislation on the management of radioactive waste or spent fuel (Article 9(3)(b)). In any case, conditions attached to the consent must be non-discriminatory, i.e., they may not be more stringent than requirements for similar shipments within that Member State.

Apart from the 2006 Shipments Directive, the new Directive 2011/70 /Euratom on the management of spent fuel and radioactive waste also (partly) applies to the transport of such waste. Its scope extends to all stages of spent fuel and radioactive waste management.[34] More specifically, Article 4.2 says:

29 Cf. Art. 1(3)(c) of the Basel Regulation.

30 Council Directive 92/3/Euratom of 3 February 1992 regarding the supervision and control of shipments of radioactive waste between Member States into and out of the Community.

31 OJ 2006, L 337/21.

32 The Convention entered into force on 18 June 2001. Since 2005 Euratom has been a party to the Convention (see Decision 2005/84/Euratom, OJ 2005, L30/10).

33 OJ 2006, L 337/21.

34 Art. 2 states: '1. This Directive shall apply to all stages of: (a) spent fuel management when the spent fuel results from civilian activities; (b) radioactive waste management, from generation to disposal, when the radioactive waste results from civilian activities. 2. This Directive shall not apply to waste from extractive industries which may be radioactive and which falls within the scope of Directive 2006/21/EC.' (above n. 17).

> (...)
> 2. Where radioactive waste or spent fuel is shipped for processing or reprocessing to a Member State or a third country, the ultimate responsibility for the safe and responsible disposal of those materials, including any waste as a by-product, shall remain with the Member State or third country from which the radioactive material was shipped.

The Member State where the radioactive waste is generated shall dispose of its radioactive waste, unless it has concluded a shipment agreement, in accordance with Article 16(2) of the 2006 Shipments Directive, with another Member State or with a third country. It has to inform the Commission prior to a shipment to a third country (but not in the case of shipment to another Member State!) of the content of such an agreement, which has to fulfil specific safety and other criteria, stipulated in the 2011 Directive.[35]

1.5 Concluding Remarks

- Will there be any improvement in public safety regarding radioactive waste management (including that of spent fuel) as a result of the new Euratom legislation? It is likely that there will be.
- Will we see better information provided as well as more effective participation in decision-making processes regarding spent nuclear fuel and radioactive waste management (including 'siting' of facilities)? I believe so because the Framework Directive promises great improvement over the situation as it was. The reference to conformity to national legislation may, however, prove to be a formidable barrier.
- Participation in the European decision-making process regarding spent fuel/radioactive waste management, notably by the EP, requires profound changes in Euratom which still does not seem to be a topic under consideration. The recent Case C-48/14 submitted to the Court by European Parliament may open up matters.
- It seems to me at least optimistic to assume that the 'self-assessment' and, in particular, the 'peer review' provision of Article 14(3) of the new Euratom Framework Directive 2011/71 are adequate to ensure a high standard of care for the management of

35 Art. 4.4.a-c, of the 2011 Directive; the agreement should be in accordance with Art. 16(2) of the 2006 Shipments Directive. Perhaps these requirements can avoid in the case of radioactive waste a disaster like the 2006 *Probo Koala* dumping of toxic waste in Côte d'Ivoire (see, for an overview of this toxic waste incident, which also led to numerous court cases, e.g., <www.marineinsight.com/misc/marine-safety/the-probo-koala-toxic-ship-incident-consequences>).

radioactive waste (and spent fuel), as the Preamble to the Directive states.[36] I wonder if leaving the industry, basically, to itself does not mean that a much too serious risk is taken in view of the potential dangers involved (notably the still unsolved matter of how to cope with this kind of waste). I believe that there is an element of irresponsibility involved in leaving such an (extremely) high spending sector (including huge public funds) accountable, based on peer review only.

- Will the new Directives be effective to prevent disastrous accidents in radioactive waste/spent fuel transport? In respect of transport to third countries, that may be, but much will depend on the active engagement of the European Commission in this respect. Regarding intra-EU transports not that much seems to have changed.

36 '(40) Peer review could serve as an excellent means of building confidence and trust in the management of radioactive waste and spent fuel in the European Union, with the aim of developing and exchanging experience and ensuring high standards, (...).'

2 Euratom Law on Radioactive Waste: *Ius Specialis*?

Miguel Sousa Ferro

2.1 Radioactive Waste and the Euratom Community/EU

Radioactive waste (*stricto sensu*) is mentioned twice in the Euratom Treaty. First, in Article 37, which obliges Member States to provide the European Commission with 'general data' relating to planned disposals of radioactive waste, to allow the issuance of an Opinion concerning potential effects on the water, soil or airspace of other Member States.[1]

The Commission frequently issues Opinions under this provision,[2] and it adopted a Recommendation to clarify the terms of its application.[3] The ECJ clarified the obligations arising from Article 37 Euratom, protecting the Commission's consultative role, making it mandatory to supply the Commission with the necessary information in time for it to issue an opinion that can still be duly taken into account by the Member State.[4] More recently, as will be further discussed below, the Court excluded radioactive waste associated to military activities from the scope of Article 37 Euratom.[5]

1 The processing of irradiated fuel is also mentioned as one of the possible areas of research to be promoted by the European Commission in accordance with Art. 9 Euratom.

2 For examples of recent Commission Opinions under Art. 37 Euratom see: Opinion of 27 March 2013 relating to the plan for the disposal of radioactive waste arising from the decommissioning of Units 1 and 2 of the Kozloduy NPP (Bulgaria) (OJ C 92/1, 28.03.2013); Opinion of 13 March 2013 relating to the plan for the disposal of radioactive waste arising from the decommissioning of the Latina Nuclear Power Plant, located at Latina (Italy) (OJ C 78/1, 16.03.2013); Opinion of 22 February 2013 relating to the plan for the disposal of radioactive waste arising from the decommissioning of the Studsvik R2 and R2-0 research reactors, located at Nyköping (Sweden) (OJ C 54/1, 23.02.2013); Opinion of 20 February 2013 relating to the plan for the disposal of radioactive waste arising from the Solid Waste Management and Storage Facility, located on the Ignalina NPP (Lithuania) (OJ C 51/1, 22.02.2013) etc.

3 Commission Recommendation 2010/635/Euratom of 11 October 2010 on the application of Art. 37 of the Euratom Treaty (OJ L 279/36, 23.10.2010). This replaced Commission Recommendation 1999/829/Euratom, which in turn followed Recommendation 91/4/Euratom, Recommendation 82/181/Euratom and the first Recommendation published in OJ 81/1893, 21.12.1960.

4 Judgment of the ECJ of 22 September 1988, *Land de Sarre* (187/87), ECR (1988) 5013.

5 Judgment of the ECJ of 12 April 2005, *Commission v. UK* (C-61/03), ECR (2005) I-2477. The *Land Wien* Case also raised an issue relating to Art. 37 Euratom, but the application was found inadmissible, see: Order of the EGC of 20 September 2011, *Land Wien v. Commission* (T-267/10) and Order of the ECJ of 12 July 2012, *Land Wien v. Commission* (C-608/11 P) (OJ C 366/21, 24.11.2012).

Second, irradiated nuclear fuel is mentioned in Article 78 Euratom, as a material subject to safeguards and, within that scope, the Commission is *inter alia* given the power to approve the techniques to be used in its chemical processing.

The Euratom Treaty does not explicitly foresee the adoption of binding rules relating, specifically and directly, to radioactive waste. Article 2(b) says that the Community should 'establish uniform safety standards to protect the health of workers and of the general public and ensure that they are applied'.

This general task is then elaborated on in Chapter 3 of Title II, relating to health and safety. Within that Chapter, Article 30 foresees the adoption of basic standards 'for the protection of the health of workers and the general public against the dangers arising from ionizing radiations'. 'Basic standards' mean 'maximum permissible doses compatible with adequate safety', 'maximum permissible levels of exposure and contamination' and 'fundamental principles governing the health surveillance of workers'. Article 31 sets out the procedure for the adoption of these basic standards. Article 32 provides for the revision and supplement of the basic standards. Article 33 further establishes the consultation of the Commission prior to the adoption of national legislation in this domain, to allow it to formulate recommendations with the goal of harmonisation.

While it might seem that such provisions do not allow for the adoption of Community rules on spent nuclear fuel or radioactive waste, the competences of the European Communities were, for many years, interpreted with a functional approach. In other words, it became settled case law of the ECJ that such competences had to be interpreted widely in light of the objectives underlying them. The Court did so, not by resorting to Article 203 Euratom, but by broadly interpreting the powers granted (especially) by Articles 2(b) and 30-32 Euratom. As stated in 1988, the ECJ considers that the provisions of Chapter 3 of Title II of the Euratom Treaty 'form a coherent whole conferring upon the Commission powers of some considerable scope in order to protect the population and the environment against the risks of nuclear contamination'.[6]

Following the *Nuclear Safety Convention* case,[7] the Commission felt confident enough, in this regard, to put forward a Nuclear Package including proposals of two Directives, relating to nuclear safety and radioactive waste management.[8] After the failure of these first proposals,[9] watered down versions of each Directive were proposed and adopted.

6 Judgment of the ECJ of 22 September 1988, *Land de Sarre* (187/87), ECR (1988) 5013, §11.

7 Judgment of the ECJ of 10 December 2002, *Commission v. Council* (C-29/99), ECR (2002) I-11221.

8 Amended proposal for a Council Directive (Euratom) laying down basic obligations and general principles on the safety of nuclear installations as well as amended proposal for a Council Directive (Euratom) on the safe management of the spent nuclear fuel and radioactive waste, 8 September 2004, COM(2004)526 final. For the 2003 version, see: COM(2003)0032 final.

9 For more on this see Sousa Ferro, M., 'The Future of the Regulation of Nuclear Safety in the EU', 2(2) (2008) *International Journal of Nuclear Law* 149, and Sousa Ferro, M., 'Directive 2009/71/Euratom: The Long Lost

Concerning radioactive waste and spent nuclear fuel, today's Law of the Euratom Community includes a significant number of relevant legal instruments (this list does not include acts which cover issues that may arise in the context of the management of radioactive waste, but which are not specific to it):

1. the Basic Safety Standards Directive[10] regulates a multitude of issues associated with radioactive waste and spent nuclear fuel management; *inter alia*, it imposes licensing and record-keeping obligations, mandatory consultation of radiation protection experts, requirements relating to disused sealed sources, provisions dealing with safety concerns associated to orphan sources, etc.;
2. the Radioactive Waste Management Directive[11] implements a European (somewhat different) version of the Joint Convention on the Safety of Spent Fuel Management and on the Safety of Radioactive Waste Management;[12]
3. the Radioactive Waste Shipments Directive[13] regulates the transboundary movement of radioactive waste from, into and between Member States;
4. shipments of radioactive waste between Member States are also governed by the Radioactive Substances Shipments Regulation;[14]

Battle against Discrimination and Protection of Sovereignty', 2(4) (2010) *International Journal of Nuclear Law* 295.

10 Council Directive 2013/59/Euratom of 5 December 2013 laying down basic safety standards for protection against the dangers arising from exposure to ionizing radiation, and repealing Directives 89/618/Euratom, 90/641/Euratom, 96/29/Euratom, 97/43/Euratom and 2003/122/Euratom (OJ L 13/1, 17.01.2014).

11 Council Directive 2011/70/Euratom of 19 July 2011 establishing a Community framework for the responsible and safe management of spent fuel and radioactive waste (OJ L 199/48, 02.08.2011).

12 To which the Euratom Community acceded, see: Commission Decision 2005/510/Euratom of 14 June 2005 concerning the accession of the European Atomic Energy Community to the 'Joint Convention on the Safety of Spent Fuel Management and on the Safety of Radioactive Waste Management' (OJ L 185/33, 16.07.2005), as well as Council Decision 2005/84/Euratom (OJ L 30/10, 03.02.2005).

13 Council Directive 2006/117/Euratom of 20 November 2006 on the supervision and control of shipments of radioactive waste and spent fuel (OJ L 337/21, 05.12.2006). Replaced: Council Directive 92/3/Euratom of 3 February 1992 (OJ L 35/24, 12.02.1992). Implemented by Commission Decision 2008/312/Euratom of 5 March 2008 establishing the standard document for the supervision and control of shipments of radioactive waste and spent fuel referred to in Council Directive 2006/117/Euratom (OJ L 107/32, 17.04.2008), which replaced Commission Decision 93/552/Euratom of 1 October 1993 (OJ L 268/83, 29.10.1993). See also: Commission Recommendation 2009/527/Euratom of 7 July 2009 for a secure and effective system of transmission of documents and information relating to the provisions of Council Directive 2006/117/Euratom (OJ L 177/5, 08.07.2009). This Directive gave rise to an infringement proceeding for lack of transposition, but the request was withdrawn before a judgment was reached, see: Order of the ECJ of 22 November 2010, *Commission v. Greece* (C-353/10).

14 Council Regulation (Euratom) 1493/93 of 8 June 1993 on shipments of radioactive substances between Member States (OJ L 148/1, 19.06.1993). See also: Communication from the Commission concerning Council Regulation (Euratom) 1493/93 on shipments of radioactive substances between Member States (OJ C 41/2, 19.02.2009), and Communication concerning Council Regulation (Euratom) 1493/93 of 8 June 1993 on shipments of radioactive substances between Member States (OJ C 40/4, 14.02.2002).

5. the regime established by the Nuclear Safety Directive[15] encompasses associated waste storage facilities;
6. in accordance with Article 78 Euratom, the Safeguards Regulation also applies to radioactive waste and establishes specific rules for the accounting, processing as well as transfer of this type of material;[16]
7. waste management issues are discussed in the framework of the European Nuclear Safety Regulators Group (ENSREG);[17] and
8. several non-binding documents have also been adopted in this area.[18]

It is also possible to find legislation adopted under the EC Treaty/TFEU which impacts on the management of radioactive waste, but in a far smaller number and always in the context of regulating broader issues, of which what would typically be Euratom issues are but a very small part. Such is the case, for example, with the:

1. EIA Directive,[19] which requires an EIA to be carried out for installations for the reprocessing of irradiated nuclear fuel, processing of irradiated nuclear fuel or high-level radioactive waste, and for long-term storage as well as final disposal of radioactive waste;

---

15 Council Directive 2009/71/Euratom of 25 June 2009 establishing a Community framework for the nuclear safety of nuclear installations (OJ L 172/18, 02.07.2009), as revised by Council Directive 2014/87/Euratom of 8 July 2014 (OJ L 219/41, 25.07.2014).

16 Commission Regulation (Euratom) 302/2005 of 8 February 2005 on the application of Euratom safeguards (OJ L 54/1, 28.02.2005). Replaced: Commission Regulation (Euratom) 3227/76, of 19 October 1976 (OJ L 363/1, 31.12.1976), amended by Commission Regulation (Euratom) 220/90 of 26 January 1990 (OJ L 22/56, 27.01.1990) and Commission Regulation (Euratom) 2130/93 of 27 July 1993 (OJ L 191/75, 31.07.1993). This, in turn, was preceded by: Commission Regulation No. 8 defining the nature and the extent of the requirements referred to in Art. 79 of the Treaty (OJ 34/651, 29.05.1959).

17 Established by Commission Decision 2007/530/Euratom of 17 July 2007 on establishing the European High Level Group on Nuclear Safety and Waste Management (OJ L 195/44, 27.07.2007). See also: Council Decision 80/237/Euratom of 18 February 1980 on the setting up of an 'ad hoc' Advisory Committee on the Reprocessing of Irradiated Nuclear Fuels (OJ L 52/9, 26.02.1980).

18 See: previously quoted Commission Recommendation 2010/635/Euratom; Commission Recommendation 2008/956/Euratom of 4 December 2008 on criteria for the export of radioactive waste and spent fuel to third countries (OJ L 338/69, 17.12.2008); Commission Recommendation 2004/2/Euratom of 18 December 2003 on standardised information on radioactive airborne and liquid discharges into the environment from nuclear power reactors and reprocessing plants in normal operation (OJ L 2/36, 06.01.2004); Commission Recommendation 99/669/EC, Euratom of 15 September 1999 on a classification system for solid radioactive waste (OJ L 265/37, 13.10.1999); Commission Recommendation 82/74/Euratom of 3 February 1982 on the storage and reprocessing of irradiated nuclear fuels (OJ L 37/36, 10.02.1982); Council Resolution of 18 February 1980 on the reprocessing of irradiated nuclear fuels (OJ C 51/4, 29.02.1980). The financial dimension of decommissioning, including spent fuel and radioactive waste management, has also been the object of several non-binding documents.

19 Directive 2011/92/EU of the European Parliament and of the Council of 13 December 2011, on the assessment of the effects of certain public and private projects on the environment (codification) (OJ L 26/1, 28.01.2012).

2. Inland Transport Safety Directive,[20] which defines safety rules for the transport of dangerous goods, including radioactive waste;
3. Extractive Industries Waste Directive,[21] which applies to waste from mining operations including naturally occurring radiation; and
4. Hazardous Substances in Electronic Equipment Directive,[22] which applies to equipment that uses ionizing radiation.

On the other hand, the EU's general Waste Directive specifically excludes radioactive waste from its scope.[23]

2.2 EURATOM AS A *LEX SPECIALIS*

It is clear that the Euratom Treaty and the provisions adopted under it constitute a *lex specialis* in relation to the TEU/TFEU as well as their secondary legislation.

The Treaties themselves refer to it, in Article 106a(3) Euratom (previously Article 305(2) EC), according to which 'the provisions of the [TEU] and of the [TFEU] shall not derogate from the provisions of this Treaty'.

This point was also settled in the case law : 'the [Euratom] Treaty constitutes (…) a lex specialis in derogation from the lex generalis represented by the EC Treaty'.[24] The same point was confirmed in the relationship between the (expired) ECSC Treaty and the TFEU.[25]

In the words of the Court, the EU legal order includes the principle that 'any general rule (…) may be limited or excluded – according to the principle that a special rule derogates

20 Directive 2008/68/EC of the European Parliament and of the Council of 24 September 2008 on the inland transport of dangerous goods (OJ L 260/13, 30.09.2008), as subsequently amended.
21 Directive 2006/21/EC of the European Parliament and of the Council of 15 March 2006 on the management of waste from extractive industries (OJ L 102/15, 11.04.2006), revised by Regulation (EC) 596/2009 (OJ L 188/14, 18.07.2009).
22 Directive 2011/65/EU of the European Parliament and of the Council of 8 June 2011 on the restriction of the use of certain hazardous substances in electrical and electronic equipment (OJ L 174/88, 01.07.2011).
23 Directive 2008/98/EC of the European Parliament and of the Council of 19 November 2008 on waste and repealing certain Directives (OJ L 312/2, 22.11.2008). See also Council Regulation (EU) 333/2011 of 31 March 2011 establishing criteria determining when certain types of scrap metal cease to be waste under Directive 2008/98/EC of the European Parliament and of the Council (OJ L 94/2, 08.04.2011), which establishes a bridge with Euratom provisions concerning radioactive scrap metal.
24 Judgment of the EGC of 6 May 2009, *Outokumpu Oyj et al v. Commission* (T-122/04), ECR (2009) II-1135, §55.
25 See, e.g.: Judgment of the EGC of 25 October 2007, *Reinforcing Bars* (T-27/03 etc.), ECR (2007) II-4331; Judgment of the EGC of 8 July 2003, *Verband der freien Rohrwerke* (T-374/00), ECR (2003) II-2275, §68; Judgment of the EGC of 24 September 2002, *Falck* (C-74/00 P etc.), ECR (2002) I-7869, §100; and Judgment of the ECJ of 15 December 1987, *Deutsche Babcock Handel* (328/85), ECR (1987) 5119, §10.

from the general rule (*lex specialis derogat legi generali*) – where there are special rules governing specific matters'.[26]

This means that one must carry out a case-by-case analysis: Only in the presence of a contradiction between provisions of both Treaties can there be a derogation or exclusion.

Therefore, the real issue is the identification of the precise consequences of this relationship of *lex specialis / lex generalis*, and of how the two are to be articulated. Looking at this issue, and at the specific solutions proposed by the Court as well as by the legislator in concrete cases, we need to assess whether the principle that Euratom is a *lex specialis* is fully translating into what we would expect from legal theory.

For the purposes of the present paper, a few issues have been chosen regarding the application of the *lex specialis* principle, having to do with: (i) the interpretation of Euratom provisions, (ii) the discussion of whether and to what extent TEU/TFEU primary as well as secondary legislation encompasses radioactive waste, and (iii) the choice of legal basis for legislation dealing with radioactive waste.

2.3 The Interpretation of Euratom Provisions

A consistent feature in a great number of cases of Nuclear Law handled by the ECJ is the affirmation – explicit or implicit – of the existence of a single EU legal order, which includes, *inter alia*, the TFEU and the Euratom Treaty. The Court has summarized this approach as follows:

> it should be noted that the founding treaties established a single legal order in which the [Euratom] Treaty constitutes, and the CS Treaty constituted until 23 July 2002, a lex specialis in derogation from the lex generalis represented by the EC Treat.[27]

This has many important practical consequences for the interpretation and application of Euratom provisions.

First, it allows the Court to refer indistinctively to its case law under the TFEU when discussing parallel provisions of the Euratom Treaty.[28] In other words, the conclusions

26 Judgment of the ECJ of 9 March 2006, *Vicenzo le Voci* (T-371/03), ECR (2005) II-957, §122.

27 Judgment of the EGC of 6 May 2009, *Outokumpu Oyj et al v. Commission* (T-122/04), ECR (2009) II-1135, §55.

28 See, e.g.: Judgment of the ECJ of 27 March 1990, *Alfredo Grifoni v. Commission* (C-308/87), ECR (1990) I-1203; Judgment of the ECJ of 29 June 1995, *Commission v. Italy* (C-135/94), ECR (1995) I-1805; Order of the ECJ of 12 January 2011, *Eriksen et al v. Commission* (C-205/10 P, C-217/10 P and C-222/10); Judgment of the EGC of 16 March 2005, *EnBW Kernkraft GmbH v. Commission* (T-283/02), ECR (2005) II-913; Judgment of the EGC of 17 September 2007, *France v. Commission* (T-240/04), ECR (2007) II-4035.

drawn from the interpretation of the provisions of the *lex generalis* tend to be automatically extended to the *lex specialis*, when interpreting identical provisions.

When extending settled case law to Article 296 TFEU and the scope of the Euratom Treaty, the Court noted that

> Article 162 Euratom, being 'substantially the same', 'must therefore be interpreted by analogy with that provision'.[29] The clarification that the recourse to TFEU case law must be made by analogy is important, but should not be overestimated since, in the very same Judgment, the Court extended to the Euratom Treaty principles clarified under the TFEU, concerning misuse of powers, without any discussion of analogy.[30]

Second, because the Court views the Euratom Treaty as 'the application, in a highly specialised field, of the legal conceptions which form the basis of the structure of the general common market',[31] it applies to matters covered by the Euratom Treaty general principles of EU law, even if they are not explicitly mentioned in it (e.g. prohibition of discrimination on the grounds of nationality or obligations of national courts when confronted with conflicts between national and EU Law), as well as TFEU case law clarifying such principles.[32]

While this is still fundamentally compatible with the *lex specialis* approach, as it corresponds to the filling of gaps by resorting to the *lex generalis*, it should be noted that it should require a teleological judgment as to the adequacy of these general principles within the specific logic as well as purposes of the Euratom Treaty, and that while such a judgment may be present in the background, it is not explained by the Court. Without losing sight of the small number of cases where this issue has been raised, the Court has shown a tendency to automatically extend to the scope of the *lex specialis* the principles developed under the *lex generalis*.

29 See: Judgment of the EGC of 25 February 1997, *Kernkraftwerke Lippe-Ems GmbH v. Commission* (T-149/94 and T-181/94), ECR (1997) II-161, §§46-47 and 144; Judgment of the EGC of 17 September 2007, *France v. Commission* (T-240/04), ECR (2007) II-4035, §30.

30 See: Judgment of the EGC of 25 February 1997, *Kernkraftwerke Lippe-Ems GmbH v. Commission* (T-149/94 and T-181/94), ECR (1997) II-161, §§53 and 149.

31 See, e.g.: Ruling of the ECJ 1/78 of 14 November 1978, ECR (1978) 2151, §15; Judgment of the EGC of 15 September 1995, *Empresa Nacional de Urânio SA v. Commission* (T-458/93 and T-523/93), ECR (1995) II-2459, §70.

32 See, e.g.: Judgment of the ECJ of 27 October 2009, *Land Oberösterreich* (C-115/08), ECR (2009) I-10265, §§88-91, 108 et seq. and 138.

Third, on a point of procedural relevance, it has allowed the Court to disregard the fact that an action was brought exclusively under the TFEU and to proceed to apply Euratom provisions to the solution of the case placed before it.[33]

All this being said, it may be argued that the principle of the single EU legal order has not always been harmoniously applied when interpreting the Euratom Treaty. There is one fundamental difference between the TFEU and the Euratom Treaty that seems to be a driver for divergent interpretations, or methods of interpretation, and that is the fact that the TFEU has been repeatedly revised as well as developed, whereas the substance of the Euratom Treaty has remained largely unchanged (even though scientific knowledge and the context in which the Treaty is applied have evolved substantially since 1957).

In 2001, Advocate General Jacobs argued that this circumstance should justify applying a different method to the interpretation of the Euratom Treaty in comparison to the TFEU.[34] Specifically, Jacobs said that while

> subsequent practice of the EU Institutions is not taken into account in the interpretation of the TFEU, it should be considered in the interpretation of the Euratom Treaty. Specifically, this Opinion meant that Articles 30 to 39 of the Euratom Treaty, and the competences therein awarded to the Community, should be interpreted in light of what Member States (through the Council) accepted to include in the Euratom Directives adopted under these provisions. While, in its Judgment in this case, the ECJ adopted a very broad interpretation of the Euratom Community's powers, as proposed by the AG, it did not seem to take up his approach on this precise point, thereby confirming the unity of the EU legal order and of the interpretative methods to be applied within it.

However, as we have already argued in greater detail,[35] the following years saw a drastic shift in the case law of the Court. In 2002, in its *Nuclear Safety Convention* case, the ECJ still appeared to be an outspoken fan of a broad interpretation of the Euratom Treaty, and a near-revolutionary supporter of the expansion of the transfer of sovereign powers from the Member States to the Community.

This was the Court as we are used to seeing it under the EC Treaty (now TFEU), using a teleological approach to champion an ever-growing scope of EU power and regulation. It was this characteristic teleological approach that the Court applied to the Euratom Treaty

33 See, e.g.: Judgment of the ECJ of 27 October 2009, *Land Oberösterreich* (C-115/08), ECR (2009) I-10265, §84; Judgment of the ECJ of 29 March 1990, *Greece v. Council* (C-62/88), ECR (1990) I-1527, §8.

34 Opinion of AG Jacobs delivered on 13 December 2001, in *Commission v. Council* (C-29/99), ECR (2002) I-11221, §§147-149.

35 See: Sousa Ferro, M., 'Nuclear Law at the European Court in the 21st Century', 3(11/12) (2012) *Revista da Concorrência e Regulação* 55.

in 2002, when it concluded that the purpose of the Treaty's rules on radiological protection 'was to ensure consistent and effective protection of the health of the general public against the dangers arising from ionising radiations, whatever their source'.[36] As a consequence of this functional approach, Euratom provisions should be applicable despite the source of the ionizing radiation. Regardless of whether radiation had a natural or non-natural source (civilian or military), the Euratom Treaty granted the Community the power to protect the health of the general public.

By mid-2005, however, the Court produced its surprising conclusion, twice confirmed since, that no provision of the Euratom Treaty applies to military activities.[37] This position has since then been reflected in the Radioactive Waste Management Directive (Article 2(1) explicitly excludes from its scope waste arising from military activities). That conclusion was reached with the vehement opposition of AG Geelhoed, and its justification is far from convincing, in both a literal as well as a teleological sense, not to mention awe-striking in light of the Court's previous Euratom case law and of its general case law under the TFEU. There is no mention in any of the case law examples prior to 2005 of such a restrictive interpretation of the scope of the Euratom Treaty.

In a Competition Law case of 2006/2007, the ECJ surfaced as protecting Member States' freedom of choice concerning the financing of the nuclear sector within their borders.[38] By contrast, in a 1982 Case, the ECJ refused to give any special protection to public undertakings in the nuclear sector.[39]

In 2007, the Court annulled a Euratom Regulation adopted by the Commission for lack of powers, all the while recognizing it was possible, in theory, for powers to be implicitly granted and their exercise necessary to give practical effect to the implemented provisions.[40] In practice, this ruling prevented the Commission from modifying the mandatory legal consequences of a notification under Article 41 Euratom.

In 2006, the Court drastically reduced the powers of the Euratom Supply Agency by excluding uranium enrichment contracts from the concept of 'supply contracts', meaning that the Agency would not have, e.g., a right of option, a right of ownership or an exclusive

36 Judgment of the ECJ of 10 December 2002, *Commission v. Council* (C-29/99), ECR (2002) I-11221, §80.

37 Judgment of the ECJ of 12 April 2005, *Commission v. UK* (C-61/03), ECR (2005) I-2477; Judgment of the ECJ of 9 March 2006, *Commission v. UK* (C-65/04), ECR (2006) I-2239; and Order of the ECJ of 12 January 2011, *Eriksen et al v. Commission* (C-205/10 P, C-217/10 P and C-222/10).

38 Judgment of the ECJ of 29 November 2007, *Stadtwerke Schwäbisch Hall GmbH et al v. Commission et al* (C-176/06 P), ECR (2007) I-170; and Judgment of the EGC of 26 January 2006, *Stadtwerke Schwäbisch Hall GmbH et al v. Commission* (T-92/02), ECR (2006) II-11.

39 Judgment of the ECJ of 6 July 1982, *France et al v. Commission* (188 to 190/80), ECR (1982) 2545, §§28 and 32.

40 Judgment of the EGC of 17 September 2007, *France v. Commission* (T-240/04), ECR (2007) II-4035.

right to conclude such contracts.[41] Free enterprise prevailed over supervision of the market by the Euratom Supply Agency.

All this is a far cry from the attitude of the Court that, for example, single-handedly prevented the disappearance of the Euratom Supply Agency and of the common supply policy, when Member States could not agree on a future for this policy, as required by the Treaty;[42] that placed the protection of the population from the dangers of ionizing radiation above the interests of the common market;[43] that walked a fine line between a possible future need for protectionist measures of Community uranium producers as well as the absence of such a need in the specific case, and ensured a broad discretionary margin for the Commission as well as the Supply Agency in the management of the common supply policy;[44] or that extended the competencies of the Euratom Community to physical protection as well as to nuclear safety, focusing on the *effet utile* of the Treaty's provisions, even though no explicit reference to these issues is found in them, and going against the wishes of the Member States.[45]

What has changed since 2002? The main event of relevance in European integration in the period that followed was the Convention on the Future of Europe, which presented its draft Constitution in July 2003. In mid-2004, this proposed Constitution was adopted by the Heads of State and Government of the EU, and was set aside following the two negative national referendums in 2005. The Constitution was eventually reshaped into the Lisbon Treaty, which finally came into force in 2009. The gruelling process of negotiations made one thing notoriously clear: There was no agreement on the future of the Euratom Treaty, and the only possible solution was not to change it in any significant way. Unlike its sibling-Treaty, the Euratom Treaty has indeed remained, in substance, unchanged since its inception.[46]

---

41 Judgment of the ECJ of 12 September 2006, *Indústrias Nucleares do Brasil* (C-123/04 and C-124/04), ECR (2006) I-7861.

42 Judgment of the ECJ of 14 December 1971, *Commission v. France* (7/71), ECR (1971) 1003.

43 Judgment of the ECJ of 25 November 1992, *Commission v. Belgium* (C-376/90), ECR (1992) I-6153.

44 Judgment of the ECJ of 16 February 1993, *Empresa Nacional de Urânio SA v. Commission* (C-107/91), ECR (1993) I-599; Judgment of the EGC of 15 September 1995, *Empresa Nacional de Urânio SA v. Commission* (T-458/93 and T-523/93), ECR (1995) II-2459; Judgment of the ECJ of 11 March 1997, *Empresa Nacional de Urânio SA v. Commission* (C-357/95 P), ECR (1997) I-1329; Judgment of the EGC of 25 February 1997, *Kernkraftwerke Lippe-Ems GmbH v. Commission* (T-149/94 and T-181/94), ECR (1997) II-161; and Judgment of the ECJ of 22 April 1999, *Kernkraftwerke Lippe-Ems GmbH v. Commission* (C-161/97 P), ECR (1999) I-2057.

45 Ruling of the ECJ 1/78 of 14 November 1978, ECR (1978) 2151; Judgment of the ECJ of 10 December 2002, *Commission v. Council* (C-29/99), ECR (2002) I-11221.

46 In short, the Amsterdam Treaty made a few substantial changes, which were either minor or had to do with the transfer of certain issues to be part of the scope of the EC Treaty. Other Treaties, such as Nice, Maastricht and Accession Treaties, also revised some provisions, but with no significant impact at the level of policy. The Lisbon Treaty mostly deleted the institutional framework provisions, ordering that the TEU's/TFEU's provisions in that regard be applied.

Specifically at the level of Nuclear Law, the 2002-2005 period was marked by the failure of the Commission's proposed Nuclear Package, which would have introduced a first set of relatively modest rules concerning nuclear safety as well as management of spent fuel and radioactive waste.[47] Significantly, the proposal of the Nuclear Package was made possible by the broad interpretation of Euratom competences put forward by the Court in the *Nuclear Safety Convention* case.[48]

This shift in the case law of the Court, as well as in particular the reversal of the Court's classical paradigm of being a champion of European integration and refusing interpretations that restrict the transfer of powers from the Member States, suggest that the ECJ has possibly grown tired of a Treaty that Member States cannot agree on revising and updating.

It has, in essence, destroyed the idea of a single EU legal order, by imposing different interpretative methods. Where is the Court's characteristic approach to interpretation that aims at guaranteeing the *effet utile* of EU provisions above all other considerations? It is clear that several Euratom objectives cannot 'be fully achieved if the applications of ionizing radiation in defence related activities are excluded from the scope of the Treaty'.[49] The conclusion that must be drawn from these cases is that, as interpreted by the Court, ensuring the safety of persons and of the environment is apparently not an overriding interest within the Euratom Treaty.[50]

The unity of the legal order has also been damaged by the creation of internal contradictions. For example, as a result of this case law, if the EU is affected by radioactive fallout from a military use of nuclear energy or ionizing radiation, the Regulations on foodstuffs contaminated by radiation will be applicable – because they were adopted (arguably, illegitimately) under the EC Treaty – but the Euratom Treaty and all of its implementing radiological protection provisions, including those relating to emergency response, will not be applicable. Clearly, this is a consequence that the Court did not foresee, nor intend, and the Court would have to find some way around it.

Without losing sight of the fact that there may still be hope that this new approach in the case law has not yet consolidated, it is important to consider what this new paradigm means for the conceptualisation of the Euratom Treaty as a *lex specialis*.

47 See: Amended proposal for a Council Directive (Euratom) laying down basic obligations and general principles on the safety of nuclear installations and amended the proposal for a Council Directive (Euratom) on the safe management of the spent nuclear fuel and radioactive waste, 8 September 2004, COM(2004)526 final. For the 2003 version, see: COM(2003)0032 final.

48 Judgment of the ECJ of 10 December 2002, *Commission v. Council* (C-29/99), ECR (2002) I-11221.

49 Andres-Ordax, B., 'Radiological Protection and Military Activities: Recent European Case-Law', in *Nuclear Inter Jura 2007 – Proceedings*, Bruylant, Brussels, 2008, p. 549.

50 On the overriding nature of this objective, see: Opinion of AG Poiares Maduro delivered on 6 April 2006, in *Indústrias Nucleares do Brasil* (C-123/04 and C-124/04), ECR (2006) I-7861, §57.

First, it means that it is not clear that the Court will be as willing as it once was to find analogies in the interpretation of both Treaties and to extend to the Euratom Treaty the application of general principles of EU Law identified under the TEU as well as TFEU.

Second, there may also be important consequences, considering the specific context and content of the recent abovementioned Judgments, in what concerns the choice of legal basis for future EU legislation in this domain.

2.4 Does TEU/TFEU Law Encompass Radioactive Waste?

There is little doubt, in my mind, that, by and large, the primary and secondary legislation of the European Union also (potentially) applies to radioactive waste, as indeed to all other substances governed by the Euratom Treaty, insofar as the issues in question are not specifically regulated by that Treaty in a non-compatible manner.

This issue has been tackled repeatedly, in the framework of different EU policies, and the solution arrived at by the EU's Institutions, as well as by the ECJ in particular, has always been in accordance with the preceding paragraph.[51] Yet, despite a non-negligible number of doctrinal studies dedicated to this issue, it remains excessively shrouded in mystery and there are still authors who dispute the extent of the applicability of certain EU policies to the nuclear sector (in the broadest sense).

One area where this dispute has been taken to great depths is EU Competition Law. It is nearly beyond dispute that it applies to the nuclear sector.[52] The European Commission and national competition authorities have repeatedly applied European competition rules (which are entirely absent from the Euratom Treaty) to the nuclear sector.

One must distinguish between the identification of a conflict between specific provisions of Euratom as well as EU Law, and the identification of a possible conflict between objectives (or policies) pursued by the two different bodies of law.

There will typically be a very limited scope for conflicts between specific provisions. The identification of a special rule of the Euratom Treaty which may exclude the applicability of EU Law in a given matter must take into account the 'wording and the broad

51 See, e.g.: Judgment of the EGC of 15 September 1995, *Empresa Nacional de Urânio SA v. Commission* (T-458/93 and T-523/93), ECR (1995) II-2459, §70 (regarding applicability of antidumping provisions); Judgment of the ECJ of 6 July 1982, *France v. Commission* (C-188/80 etc.), ECR (1982) 2545, §§29 and 32; Opinion of the ECJ of 15 November 1994, *WTO* (1/94), ECR (1994) I-5267, §24; and Judgment of the EGC of 26 January 2006, *Stadtwerke Schwäbisch Hall GmbH et al v. Commission* (T-92/02), ECR (2006) II-11 (regarding State aid rules and rules for the conclusion of external trade agreements).

52 For more on this issue, see: Sousa Ferro, M., 'Competition Law and the Nuclear Sector: An EU Outlook', 85 (2010/2) *Nuclear Law Bulletin* 13, and the works quoted therein.

logic'[53] or the 'spirit and purpose of the rule'[54] in question. Furthermore, one should keep in mind that, under general principles of EU Law, while attempting to identify Euratom provisions which derogate from EU Law, 'the terms used to delimit [the Euratom Treaty's] scope must be given a strict interpretation'.[55]

No general derogation (relation of *lex specialis*) exists if the different provisions can be found to be complementary, as pursuing different objectives without excluding or annulling each other.[56] In other words, when different matters are being regulated, in the absence of an abstract and necessary contradiction, independent of specific circumstances, the *lex specialis* principle is not called into play, and instead we are faced with a – fairly common – situation of two sets of rules pursuing different objectives in the same situation.

In what concerns the rules on radioactive waste management and its transport, we have been unable to identify any example of contradictions between specific Euratom/EU provisions. The practice of the EU legislator confirms the interpretation herein put forward, to the extent that it has considered applicable to radioactive waste management EU rules on EIAs, transport safety and hazardous substances in electronic equipment (see above).

As for policy conflicts, it is not easy to identify (even hypothetically) examples of contradictions at the level of Treaty objectives, in the field of radioactive waste management and transport. The Euratom and EU Treaties reveal the same balance of concerns in addressing these substances, guided primarily by safety as well as the protection of the health of workers, the general public and the environment. Euratom also stresses security interests, associated to physical protection (Article 2(e)), but these do not seem to hinder the pursuit of other European objectives.

The only possible clash between objectives we could identify does not derive from contrasting the Euratom Treaty to the EU Treaties, but rather is an internal conflict present especially in the TFEU – the pursuit of internal market objectives v the pursuit of safety, health and environmental concerns. As an example, it is settled case law that environmental concerns allow for, and require, a derogation of the internal market objective, to the extent that waste should, in principle, be treated as closely as possible to its source.

Even if an abstract potential conflict were to be identified, it is our belief, based on our analysis of the case law, that a general Treaty objective (without being associated to a specific provision) may not be used as grounds for the inapplicability of EU Law. In such cases,

53 Judgment of the EGC of 28 April 2004, *Sunrider Corp* (T-124/02 etc.), ECR (2004) II-1149, §38.

54 Opinion of AG Trstenjak in *Schneider* (C-285/06), ECR (2008) I-1501, §73. See also Judgment of the ECJ of 16 January 2003, *Libor Cipra* (C-439/01), ECR (2003) I-745, §35; and Judgment of the ECJ of 13 December 2001, *Heininger* (C-481/99), ECR (2001) I-9945, §§37-40.

55 See, by analogy: Judgment of the EGC of 5 June 2001, *ESF* (T-6/99), ECR (2001) II-1523, §102; and Judgment of the EGC of 7 December 1999, *Interporc* (T-92/98), ECR (1999) II-3521, §40.

56 See: Judgment of the ECJ of 20 September 2001, *HJ Banks* (C-390/98), ECR (2001) I-6117, §83; Judgment of the EGC of 12 October 2000, *JT's Corporation* (T-123/99), ECR (2000) II-3269, §50; Judgment of the ECJ of 20 March 1957, *Ruhr Coal* (2/56), ECR (1957) 9.

the conclusion must not be that EU Law is not applicable, but that (just as in the case of opposing constitutional principles in the internal legal orders of the Member States) one principle of EU/Euratom Law must be given precedence over the other, on a case-by-case basis. This is the classic *Wouters* situation, relating to overriding objectives.[57]

In theory, therefore, it is possible that there may be a situation where the pursuit of an objective of the Euratom Treaty relating to radioactive waste may justify a derogation from the application of EU rules relating to the pursuit of a different objective. It is quite likely, however, that such a situation would also see some objectives of the EU Treaties themselves (e.g. environmental protection) on the same side as the Euratom objective in question. An assessment would have to be made concerning whether the pursuit of the first objectives would bring greater benefits (to the EU, the public, etc.) than the pursuit of the second objectives. It should also be noted that it cannot be excluded that a particularly important EU general objective may override a Euratom objective in a specific case.

2.5 Euratom as a Legal Basis for Legislation

This section will address the impact of the Euratom Treaty's status as *lex specialis* when it comes to choosing a legal basis for European legislation.

Generally, under the case law of the ECJ, if the purposes and content of a legislative act fall within two different provisions of the Treaties that award the Union/Community the competence to act, then one must determine which the primary goal is . If neither goal can be singled out as the primary one, then the act must be adopted with a dual legal basis.[58] However, the use of a dual legal basis is not possible when the provisions in question require the adoption of different and irreconcilable legislative procedures.[59] In such cases, it is still necessary to decide which provision takes precedence.

Virtually all rules adopted under the Euratom Treaty also have an impact on matters governed by the TFEU, especially on the functioning of the internal market and the pro-

57 In its Judgment of 19 February 2002, *Wouters* (C-309/99), ECR (2002) I-1577), the ECJ was asked whether EU Competition Law prevented the Dutch Bar Council from prohibiting Dutch lawyers from entering into partnerships with non-lawyers (e.g. accountants). The Court considered that such a prohibition was a decision of an association of undertakings, which restricted competition. Typically, therefore, it would be prohibited by Art. 101(1) of the TFEU, and could only be allowed if it met the requisites of Art. 101(3). However, rather than applying these requisites, the Court found that the decision was not illegal, because it was necessary for the pursuit of overriding objectives (in this case, ensuring that the ultimate consumers of legal services and the sound administration of justice were provided with the necessary guarantees in relation to integrity as well as experience). In other words, the Court recognised that policy objectives outside the sphere of competition policy can occasionally justify exemptions from Competition Law.

58 As is settled case law since the Judgment of the ECJ of 27 September 1988, *Commission v. Council* (165/87), ECR (1988) 5545.

59 As expressed since the Judgment of the ECJ of 11 June 1991, *Commission v. Council* (C-300/89), ECR (1991) I-2867.

tection of public health as well as of the environment. Conversely, some issues regulated under the TFEU have an impact on matters governed by the Euratom Treaty, specifically (in what is most relevant for the present paper) on radioactive waste management and transport.

Given that the competence and procedure for the adoption of rules under the Euratom Treaty is usually different from those of the TFEU, it can be very important to determine the correct legal basis for legislation that falls into either of those two categories. Under Articles 31-32 Euratom, for example, legislation is adopted by the Council, by qualified majority, after consulting the European Parliament. This legal basis therefore deprives the European Parliament of the co-legislator role that it has gained in the greater number of matters governed by the TFEU.

In practice, there is no legislation adopted with a dual legal basis from both Treaties in what concerns radioactive waste. It is now, indeed, very rare to find legislation with such a dual legal basis in any issue. One notable category of exception is financial Regulations, such as Regulation (EU, Euratom) 966/2012,[60] adopted under Article 322 TFEU and Article 106a Euratom (which specifically orders the application of that TFEU provision within the scope of Euratom).

Almost all the legislation adopted under Euratom (alone) relating to radioactive waste (mentioned above in Section 2.1) was adopted under Articles 31 and 32 Euratom.[61] Since the legal basis of these acts was not challenged (e.g. by the European Parliament, with two notable exceptions), this means that the Institutions believed (and I would personally agree) that the primary purpose of those legal acts was to set out basic safety standards and to supplement them, so as to protect workers as well as the general public from the dangers of ionizing radiation.

This approach is confirmed by the only case, so far decided, where a Euratom legal basis was challenged before the ECJ.[62] In this case, the European Parliament argued that Regulation 87/3954/Euratom, relating to radioactive contamination of foodstuffs following emergencies, should have been adopted under Article 95 EC (internal market). The practical consequence would have been that the Council would have had (back then) to 'cooperate' with the Parliament, rather than merely 'consult' it. The Court concluded that the purpose of that Regulation was 'to protect the population against the dangers arising

60 Regulation (EU, Euratom) 966/2012 of the European Parliament and of the Council of 25 October 2012 on the financial rules applicable to the general budget of the Union and repealing Council Regulation (EC, Euratom) No. 1605/2002 (OJ L 298/1, 26.10.2012).

61 The only exception is Regulation (Euratom) 302/2005, relating to safeguards, adopted under Arts. 77 to 79 and 81 Euratom.

62 Judgments of the ECJ of 22 May 1990 and of 4 October 1991, *European Parliament v. Council* (C-70/88), ECR (1990) I-2041 and ECR (1991) I-4529.

from foodstuffs and feedingstuffs which have undergone radioactive contamination',[63] and that it was therefore accurately adopted exclusively under Article 31 Euratom.

More recently, the European Parliament has asked the Court to find that a Directive aimed at preventing radioactive contamination of water intended for human consumption should have been adopted under the TFEU, rather than the Euratom Treaty, meaning that the EP should have been co-legislator.[64]

The issue of choice of legal basis might seem relatively peaceful, but the recent case law of the ECJ that has excluded military activities from the scope of the Euratom Treaty has created an unnecessary level of doubts and legal uncertainty. While this case law has not yet consolidated and still needs to be put to the test in several fields, the fact remains that it can now be argued that none of the legislation adopted under Euratom applies to military activities. As a result, management and transport of radioactive waste generated in the framework of military activities is arguably not subject to any of the obligations arising from that legislation.

This simply does not make sense, given that the Court has always interpreted the scope of Community competencies by resorting to a functional approach, and the encompassing of military waste is essential to achieve the Treaty's objectives of protecting workers as well as the general public from the dangers of ionizing radiation. As the Court itself noted in its earlier case law: 'the purpose of the articles referred to is to ensure the consistent and effective protection of the health of the general public against the dangers arising from ionizing radiations, whatever their source and whatever the categories of persons exposed to such radiations'.[65] In other words, the Court previously endorsed the use of Article 31 Euratom as the correct basis for the adoption of any binding provisions aimed primarily at protecting workers and the public from the dangers of ionizing radiation.

Furthermore, the Court's new position creates logical conundrums in the choice of legal basis. It was not the Court's intention that military activities with ionizing radiation go unregulated, falling into a vacuum in the EU legal order. The Court actually seemed to suggest that the TFEU be used to adopt such rules, so that they could then encompass military activities, actively promoting a change in the current legal basis of radiological

63 Judgment of the ECJ of 4 October 1991, *European Parliament v. Council* (C-70/88), ECR (1991) I-4529, §12.

64 See: Action brought on 30 January 2014, *European Parliament v. Council* (C-48/14) (OJ 2014/C 102/30), seeking annulment of Council Directive 2013/51/Euratom of 22 October 2013 laying down requirements for the protection of the health of the general public with regard to radioactive substances in water intended for human consumption (OJ L 296/12, 07.11.2013). The EP has argued that the measures foreseen in this Directive 'fall within the responsibilities of the European Union in the field of the protection of the environment, referred to in Article 192 TFEU'. It has also raised an infringement of the principle of loyal cooperation between the Institutions, and the argument that 'the contested directive undermines legal certainty in so far as it establishes rules of review and analysis which duplicate those already in force under Directive 98/83/EC'.

65 Idem, §§13-14.

protection rules. As we have noted, this may very well be rooted in the Court's dissatisfaction with the lack of reform of the Euratom Treaty, but its consequences may not have been entirely weighed.

If the EU Institutions wished the Radioactive Waste Management Directive, or the Radioactive Waste Shipments Directive, for example, to apply to military activities, could they adopt them with a TFEU legal basis, or with a dual Euratom/EU legal basis (e.g. under Article 192 TFEU) in order to make that possible, in light of this case law? I do not believe so.

Quite simply, these acts clearly have as their primary goal protection from the dangers of ionizing radiation. The pursuit of that goal falls under the purview of the *lex specialis* Euratom. The Court's own case law therefore prevents it from allowing the use of a different legal basis. That is certainly the case if one were to try to use the TFEU as a single legal basis. It would also not be possible to use it cumulatively, not only because it would still violate the case law on the primary purpose of the act in question, but also because the two legislative procedures to be applied would not be compatible. Even if one were to conclude (with some imagination) that the act in question pursued a truly dual goal, falling under both Treaties, the incompatibility between legislative procedures would force the choice of a single legal basis, and the Court has stated that, given that environmental protection is a horizontal objective to be included in all EU policies, it should not be given preference as a legal basis in such cases.

The only way to avoid such problems would be to adopt entirely separate legal instruments for radiological protection within the military sphere, with the TFEU as an exclusive legal basis. This approach would, however, likely be challenged. It would require concluding that Member States transferred their sovereign power to regulate radiological protection in the military to the EU, allowing such legislation to be adopted jointly by the European Parliament and the Council, whereas States clearly negotiated the transfer of sovereignty in this field in the special framework of the Euratom Treaty. Thus, such an approach is likely to raise the argument that if Member States did not transfer the competency to regulate radiological protection in military activities to the Euratom Community, they certainly did not intend to do so under the TFEU.

Another sensitive issue is the legitimacy of the use of the TFEU as an exclusive legal basis in several acts that influence the management and transport of radioactive waste. If the Member States have instituted a *lex specialis* for activities involving ionizing radiation, and have agreed on different legislative procedures within the scope of that *lex specialis*, why is it legitimate for the European Union, using the ordinary legislative procedure, to regulate environmental impact assessments for radioactive waste management facilities (EIA Directive, adopted under Article 192(1) TFEU), or to define the safety rules for the transport of radioactive waste (Inland Transport Safety Directive, adopted under current Article 91 TFEU), or to regulate protection from ionizing radiation in the context of mining operations (Extractive Industries Waste Directive, adopted under current Article 192(1) TFEU)?

2.6 Some Final Thoughts

It's true that the EU and Euratom have taken steps about safety in the management as well as transport of radioactive waste. If you really boil it down and pay attention to the fine print, there's a lot of substance still missing from European legislation on these issues. Member States are recognized a broad discretionary margin that fits oddly with the level of potential cross-border effects and of impact on citizens of other Member States, not to mention with the EU's as well as Euratom's obligation to ensure the radiological protection of the population of each Member State, irrespective of cross-border effects.

Something is rotten in the kingdom of Euratom.

In part, (some) Member States are to blame. There is a blocking minority, made up of old Member States with nuclear power plants, which is preventing the European Union and the Euratom Community from achieving the goal of ensuring protection against the dangers of ionizing radiation. This is one area where the subsidiarity principle clearly calls for EU action, given the notorious transboundary effects that may be involved. These Member States don't want the EU to legislate, and so it doesn't.

The two recent Directives (the 'possible' compromise that came after the failure of the initial proposal of the Nuclear Package), may do more harm than good, as they create the appearance of Community harmonization, without actually imposing specific substantive obligations, nor truly ensuring safe management. The appearance of harmonization is dangerous, because it can lead – as it has already – the Court to decide that Member States must automatically recognize standards set by neighbouring States. The fact that, in some cases, these Directives do not even go as far as existing International Law is a tell-tale sign that something is off.

This situation is made worse by the limited role granted to the European Parliament whenever Euratom legislation is at stake (the EP is merely consulted). It is surprising that the EP has remained so passive in relation to this field. Its most recent incursion into the topic, before the ECJ, has been limited to the field of radiological protection in water intended for human consumption, when there are many other areas that would arguably raise more concerns that are significant. Perhaps it is the highly sensitive political nature of it that has prevented the EP from using recent ECJ case law to call for more action on protection against ionizing radiation to be taken on the basis, not of Euratom, but of the TFEU. Clearly, this is one battle that would be worth fighting – at the very least, it would allow confronting the ECJ with the reality of the consequences of its recent case law that has interpreted the scope of Euratom restrictively.

This case law is the second part of the legal tragedy that has been quietly unfolding. Aside from the fact that the Court has reversed its previous case law and the general principles for the interpretation of the EU, without much explanation as to the reasons why it has abandoned its classic functional approach, what is most impressive is that it has seemingly developed an isolated view of Euratom, that completely ignores general principles of EU Law or international obligations that bind the Euratom Community and abandons the idea of the single EU legal order.

As is discussed in other chapters in this book, EU Law includes obligations of environmental protection that are binding to all Member States and to the Community, even in the scope of military activities. Human rights – both under the ECHR and in the EU legal order, as general principles of law – are also known to include a right to health as well as environmental protection, which cannot be automatically excluded by military activities (certainly not without applying a test of proportionality between the interests at stake). International Law includes duties of due care that translate into obligations of environmental protection, even in time of war.

In light of all this, how can it be argued that Euratom does not apply – *tout court* – to military activities, and that the Euratom Community as well as the EU have nothing to say about the endangerment of the health of the population or of the environment, if the source is of a military nature?

As public awareness on these issues is not likely to change, our only hope is for the Court to be confronted with a case that points out the gaps opened by its case law and how such gaps irreparably endanger the accomplishment of the goals of both Treaties. Perhaps then, it will fine-tune its approach and return to a functional interpretation of this dusty, but still crucial Treaty.

3 Radioactive Waste Management in the Jurisprudence of the European Court of Justice

Stefano Silingardi

3.1 Introduction

In the last decades there have not been many cases in which the European Court of Justice (ECJ) or the European General Court (EGC, i.e. the former Court of First Instance) dealt with substantial Euratom Law. According to the most recent analysis, there have been nearly fifty rulings of the ECJ and the EGC on Nuclear Law issues, the first dating back to 1971 (13 years after the entry into force of the EEC and Euratom Treaties).[1] If we consider the distribution of these rulings over the years, we might note a significant tendency for increase of litigation before the European Courts in the period between 1990 and 2011, with only seven judgements and two orders issued until 1990. However, in the last few years, the Euratom seems to have lost its allure among European litigators, the last ruling on nuclear issues being the Order of the ECJ of 12 January 2011 in the *Thule Nuclear Accident* cases.[2] Moreover, most of these cases have been described as 'similar and rather unspectacular actions initiated by the European Commission for declarations of a failure of a Member State to fulfil its obligations under certain EURATOM directives';[3] thus meaning that only few of these rulings tackle EURATOM's substantial issues.

---

1 See: Sousa Ferro M., 'Nuclear Law at the European Court in the 21st Century', *Revista de Concorrência e Regulação* (55), 2012, p. 56, Wolf S., 'EURATOM, the European Court of Justice, and the Limits of Nuclear Integration in Europe', *German Law Journal* (12), 2011, p. 1639.

2 Order of the ECJ of 12 January 2011, *Eriksen et al v. Commission* (C-205/10 P, C-217/10 P and C-222/10), ECR 2011 I-00001. More recently, see the Action brought on 30 January 2014 *European Parliament v. Council* (C-48/14) (OJ 2014/C 102/30), seeking annulment of Council Directive 2013/51/Euratom of 22 October 2013 laying down requirements for the protection of the health of the general public with regard to radioactive substances in water intended for human consumption (OJ L 296/12). In its claim, the European Parliament (EP) argued that 'the choice of legal basis made by the Council is mistaken, on the ground that the measures covered by the contested directive fall within the responsibilities of the European Union in the field of the protection of the environment, referred to in Article 192 TFEU'. Furthermore, the EP claimed that, on the one hand, 'the contested directive undermines legal certainty in so far as it establishes rules of review and analysis which duplicate those already in force under Directive 98/83/EC', and on the other hand, 'that by adopting the contested directive, the Council infringed the principle of loyal cooperation between the institutions, referred to in Article 13(2) TEU'. The case is still in progress.

3 See: Wolf S., above n. 1, p. 1639.

Notwithstanding the low number of judgments, it is, first and foremost, important to draw attention to the fact that the ECJ has a jurisprudence on radioactive waste management and nuclear safety. This is not an obvious fact, if we consider that all the international conventions on nuclear safety and radioactive waste management merely rely on peer review mechanisms (i.e. a system which creates incentives for the contracting Parties to comply).[4]

Thus, on the one hand, the EU Directives on radioactive waste management largely replicate what already exists on the international level, i.e. that the obligations are 'imported' from the international conventions into the EU legal framework.[5] On the other hand, these Directives make the 'international' obligations enforceable, i.e. through the infringement procedure, non-complying States are exposed to sanctions and litigation in the ECJ (Article 258 TFEU).

3.2 The Definition of 'Waste' and the Unresolved Problem of What 'Radioactive Waste' Means

The concept of 'radioactive waste' has not been dealt with by the ECJ, nor has the ECJ ever tried to clarify the definition of 'radioactive waste' encompassed in Article 3 paragraph 7 of the 2011/70 Council Directive.[6] This silence might seem surprising, especially if we

4 See: Södersten A., 'The EU and Nuclear Safety: Challenges Old and New', Swedish Institute for European Policy Studies (SIPES), European Policy Analysis 2012, no. 10, p. 8, <www.sipes.se>.

5 The reference regards three Directives: the Council Directive 2006/117/Euratom of 20 November 2006 on the supervision and control of shipments of radioactive waste and spent fuel, OJ L 337/21, the Council Directive 2009/71/Euratom of 25 June 2009 establishing a Community Framework for the Nuclear Safety of Nuclear Installations, OJ L 172/18, as revised by Council Directive 2014/87/Euratom of 8 July 2014, OJ L 219/41, and the Council Directive 2001/70/Euratom of 19 July 2011 establishing a Community Framework for the Responsible and Safe Management of Spent Fuel and Radioactive Waste, OJ L 199/48. Each of these Directives has been modelled on international conventions, that are, respectively, the Basel Convention on the Control of Transboundary Movement of Hazardous Waste and their Disposal, Basel, 22 March 1989, enforced 5 May 1992, United Nations, *Treaty Series*, vol. 1673, p. 57 (see also <www.basel.int>), the IAEA Convention on Nuclear Safety, Vienna, 17 June 1994, enforced 24 October 1996, 33 I.L.M. 1514 (1994), and, finally, the IAEA Joint Convention on the Safety of Spent Fuel Management and on the Safety of Radioactive Waste Management, Vienna, 5 September 1997, enforced 18 June 2011, 36 I.L.M. 1436 (1997) (see also: <www.iaea.org/Publications/Documents/Conventions/jointconv.html>). For more on the three Directives and their relationship with the international conventions mentioned above, see also Chapter 1 by Post in this volume.

6 See: Council Directive 2001/70/Euratom of 19 July 2011, above n. 5, Art. 3 para. 7: '"radioactive waste" means radioactive material in gaseous, liquid or solid form for which *no further use is foreseen or considered* by the Member State or by a legal or natural person whose decision is accepted by the Member State, and which is regulated as radioactive waste by a competent regulatory authority under the legislative and regulatory framework of the Member State' (italics added). For more on this, see also Chapter 1 by Post in this volume.

compare it with the vast and somehow confusing body of ECJ case law about the definition of 'waste' in general.

To better understand the reasons which stand behind this silence, it is important to give a brief overview of the ECJ's jurisprudence on the definition of 'waste' in general, both because it is a significant case law and it gave rise to several (not only juridical) consequences.

At the beginning of the 1990, in *Walloon Waste*, the Court said the following about the nature of waste:

> With respect to the environment, it is important to note that waste is matter of special kind. Accumulation of waste, even before it becomes a health hazard, constitutes a danger to the environment, regard being had in particular to the limited capacity of each region or locality for waste reception.[7]

From this date on, much of the Court's reasoning about 'waste' in general has been informed by this special status of waste as a latent risk to the environment. As it has been noted by some commentators, 'seeing waste as a problem rather than a potential resource has been the predominant approach of the Court'.[8]

In *Euro Tombesi* the Court noted that:

> The concept of 'waste' (...) is not to be understood as excluding substances and objects which are capable of economic reutilization, even if the materials in question may be the subject of a transaction or quoted on public or private commercial lists.[9]

7 Judgement of the Court of 9 July 1992, *Commission v. Belgium (Walloon Waste)* (C-2/90), ECR 1992 I-04431, para. 30. See: Fischer E., Lange B., Scotford E., *Environmental Law*, Oxford U. Press, 2013, p. 671. The earliest of the European Court of Justice key cases on defining waste in general was *Vessoso and Zanetti*, 28 March 1990, where the Court explained another main element: 'The concept of waste... is not to be understood as excluding substances and objects which are capable of economic re-utilization'. Judgement of the Court (First Chamber) of 28 March 1990, *Criminal Proceeding against Vessoso and Zanetti* (Joined Cases C-206/88 and C-207/88), ECR 1990 I-01461. This case concerned the original 1975 incarnation of the Waste Framework Directive, which was less focused on the recycling, recovery as well as prevention of waste, and whose definition was only limited to 'any substance or object which the holder disposes of or is required to dispose of pursuant to the provisions of national law in force', with no reference to 'intention' to dispose of. The decision in *Vessoso* was upheld even after the change of wording of the definition of waste with the 1991 amendments to the original Waste Framework Directive (dispose became 'discard', and 'intention to discard' was also included).

8 See: Fischer E., Lange B., Scotford E., above n. 7, p. 671.

9 Judgement of the Court (Sixth Chamber) of 25 June 1997, *Criminal Proceedings against Euro Tombesi and Adino Tombesi et al* (Joined Cases C-304/94, C-330/94, C-342/94 and C-224/95), ERC 1997 I-03561, para. 47.

The line of this reasoning was reinforced in a subsequent 1997 case, *Inter-Environnement Wallonie v. Region Wallonne*, in which the Court ruled that 'a substance is not excluded from the definition of waste by the mere fact that it directly or indirectly forms an integral part of an industrial production process'.[10]

With these rulings, it is clear that the ECJ intended to broaden the concept of 'waste' as far as possible. The reason of this interpretative option is not difficult to understand if we consider that according to the preamble to Directive 75/442, the Court noted that:

> the essential objective of all provisions relating to waste disposal must be the protection of human health and environment against harmful effects caused by the collection, transport, treatment, storage and tipping of waste.[11]

It is therefore clear that in the ECJ's view the concept of 'waste' cannot be interpreted restrictively and that not only substances as well as objects which are capable of economic reutilization, but also substances and objects which are capable of being recovered in an environmentally responsible manner (such as fuel for example) must be included in the concept of 'waste'. This way, the Court finally gave an answer to the unresolved issue of what 'discard' means, the idea being that, contrary to its ordinary, common-sense meaning, this concept did not merely embrace the concept of disposal of the relevant material, but also its recovery and recycling.[12]

The broad approach used by the ECJ regarding the concept of 'waste' continued in the next years,[13] both attracting consternation from most commentators and producing a

10 Judgement of the Court of 18 December 1997, *Inter-Envirronement Wallonie ASBL v. Région Wallone* (C-129/96), ERC 1997 I-07411, para. 23.

11 Judgement of the Court (Fifth Chamber) of 15 June 2000, *Arco Chemie Nederland Ltd v. Minister van Volkshuisvesting, Ruimtelijke Ordening en Milierbeher et al* (Joined Cases C-418/97 and C-419/97), ERC 2000 I-04475, para. 38.

12 See: *Arco Chemie*, above n. 11, para. 65. The Court went on, recognizing that whether a substance is 'in fact waste within the meaning of the Directive must be determined in the light of all circumstances, regard being had to the aim of the directive and the need to ensure that its effectiveness is not undermined'. The Court gave also a list of indicators that might constitute evidence of acts of discarding, such as, for example: 'where the substance used is a production residue, that is in itself a product not in itself sought for use as fuel' (para. 84), 'the fact that the substance is a residue for which no use other than disposal can be envisaged' (para. 86), 'where the substance is a residue whose composition is not suitable for the use made of it or where special precautions must be taken when it is used owing to the environmentally hazardous nature of its composition' (para. 87).

13 See: Judgement of the Court (Second Chamber) of 7 September 2004, *Criminal Proceedings against Paul Van de Walle et al* (C-1/03), ECR I-07613, para. 61, where the Court held that: 'hydrocarbons which are unintentionally spilled and cause soil and groundwater contamination are waste (...) The same is true for soil contaminated by hydrocarbons, even if it has not been excavated'. See also: Judgement of the Court (Second Chamber) of 11 November 2004, *Criminal Proceedings against Antonio Niselli* (C-457/02), ECR 2004 I-10853, para. 53: 'the meaning of 'waste' for the purposes of the first subparagraph of Art. 1(a) of Directive 75/442 is not to be interpreted as excluding all production or consumption residues which can be or are reused in a cycle of production or consumption, either without prior treatment and without harm

depressive effect on the European industry, mainly because it was practically impossible to characterize a material that was in any way left over from an industrial manufacturing or extractive process (even though useful as well as of commercial value and purpose) as anything other than 'waste'.[14]

All these rulings formed the basis for what became the simple general definition of 'waste' embraced in the 2008 Waste Framework Directive.[15] This Directive explicitly excludes from its scope radioactive waste (Article 2c), and the Court, probably because of political constraints, decided not to intervene on this subject as well as never tried to clarify the definition of 'radioactive waste' encompassed in Article 3 paragraph 7 of the 2011/70 Council Directive.

3.3 The ECJ's Radioactive Waste Case Law

In this section, we will focus on the analysis of three well-known ECJ cases on radioactive waste and nuclear safety, which, for different reasons, seem relevant, both because they can open interesting lines of reflection about the validity of the Court's role in the specific area of nuclear safety, and because what we can obtain from these decisions is useful

---

to the environment, or after undergoing prior treatment without, however, requiring a recovery operation within the meaning of Annex II B to that directive'. In a second line of Judgements, which formed the basis for Art. 5 of the Directive 2008/98/EC of the European Parliament and of the Council of 19 November 2008 on waste and repealing certain Directives, OJ L 312, the ECJ ruled that there are two circumstances in which materials that are not the primary product of an industrial manufacturing or extractive process may nonetheless be non-waste. As an example of the first circumstance, see Order of the Court (Third Chamber) of 15 January 2004, *Criminal Proceedings against Marco Antionio Saetti et al* (C-235/02), ERC 2004 I-01005, where the Court held that a non-primary product (petroleum coke, a carbon-based material produced in the refining of crude oil) is nonetheless a secondary product rather than a production residue when it is deliberately produced by technical choice as part of the primary production process. The second instance is where the material is a production residue produced unintentionally or as an inevitable consequence of the main production process, but it has characteristics that render it a by-product and not waste. See Judgement of the Court (Sixth Chamber) of 11 September 2003, *AvestaPolarit Chrome Oy* (C-114/01), ERC 2003 I-08725, and Judgement of the Court (Third Chamber) of 8 September 2005, *Commission of the European Communities v. Kingdom of Spain (The Spanish Pig Manure* case*)* (C-416/02), ERC 2005 I-07487. For more about this topic, see: Waite A., 'The Definition of Waste: The Riddle in the Sands', Frieden in Freiheit. Peace in Liberty. Paix en liberte. Festschrift für Michael Bothe zum 70. Geburtstag (eds Fischer-Lescano A., Gasser H.-P., Marauhn T., Ronzitti N.), Nomos – Dike, 2008, pp. 787 et seq.

14 See: Brown V., 'The 'End of Waste' under EU Law', Natural Resources and Environment, American Bar Association, (28, No. 3), 2014, p. 41. See also: Court of Appeal, Civil Division, Judgement of 28 June 2007, *R (on the Application of OSS Group Ltd) v. Environment Agency and Others* [2007] EWCA Civ 611, para. 55, in which Carnwath LJ stated that 'a search for logical coherence in the [ECJ] case-law is probably doomed to failure'. According to the judge, the European Law as interpreted by the ECJ made it impossible to provide a definitive ruling. He pointed to the logical incoherence of ECJ case law on the topic, with a fundamental problem identified as the Court's professed adherence to the definition in Directive 2006/12 Art. 1(a), even where it could be of no practical relevance.

15 See: Directive 2008/98/EC, above n. 13, Art. 3 para. 1, '"waste" means any substance or object which the holder discards or intends or is required to discard'.

information about the protection of EU citizens' rights in such a sensitive as well as substantial area.

3.3.1 Commission v. Council (2002): *The Definition of a Proper Legal Basis*

The case *Commission v. Council of the European Union* (so-called *Nuclear Safety Convention* case)[16] concerns the application for annulment in part of the Council Decision of 7 December 1998 approving the accession of the European Atomic Energy Community to the 1994 Nuclear Safety Convention.[17] According to the Commission, the Council sought to establish that the Community's competence in the fields covered by the Convention does not extend to the fields covered by Articles 1 to 5, 7, 14, 16(1) and (3) and 17 to 19 of the Convention. These are fundamental provisions to the Convention, which cover the authorisation system applicable to the construction and operation of nuclear power plants, assessment as well as verification of safety, emergency preparedness, siting, design, construction and operation of power plants respectively.

The Court agreed with the Commission, concluding that Euratom competencies extend to all these provisions (with the only exception of the introductory articles) and therefore ruled that the declaration of the Council must be annulled in so far as some of these provisions (Articles 7, 14, 16(1) and (3) and 17 to 19) are not referred to therein.

What is really important is the reasoning through which the Court delivered its judgement. At the core of the ECJ's ruling stands the conviction that Euratom possesses competencies relating not only to the 'traditionally' recognised *radiation protection aspects*, but also to *different aspects of nuclear safety*.

At first, the Court refers to its judgement in the landmark case *Saarland and Others* of 1988.[18] In this case, the Court held that the provisions of the chapter of the Euratom Treaty entitled 'Health and Safety' form a coherent whole conferring on the Commission powers

16 Judgement of 10 December 2002, *Commission v. Council of the European Union* (C-29/99), ECR 2002 I-11221.

17 Council Decision of 7 December 1998, approving the accession of the EAEC to the Nuclear Safety Convention. The relevant part of the declaration was worded as follows: 'The Community declares that Articles 15 and 16(2) of the Convention apply to it. Articles 1 to 5, Article 7(1), Article 14(ii) and Articles 20 to 35 also apply to it only in so far as the fields covered by Articles 15 and 16(2) are concerned. The Community possesses competence, shared with the abovementioned Member States, in the fields covered by Articles 15 and 16(2) of the Convention as provided for by the Treaty establishing the European Atomic Energy Community in Article 2(b) and the relevant articles of Title II, Chapter 3 Health and safety.' By the time of this Judgement, the IAEA Nuclear Safety Convention, above n. 6, had been ratified by all the Member States and by the Euratom Community. See: Commission Decision 1998/819/Euratom of 16 November 1999, concerning the accession to the 1994 Convention on Nuclear Safety by the European Atomic Energy Community, OJ L 318/20. The Convention entered into force for the Euratom Community on 30 April 2000.

18 Judgement of the Court of 22 September 1988, *Land de Sarre et al v. Ministre de l'Industrie, des P et T et du Tourism et al* (C-187/87), ECR 1988 05013.

of some considerable scope in order to protect the population and the environment against the risks of nuclear contamination. In the light of the purpose of Article 37 of the Euratom Treaty, which is to forestall any possibility of radioactive contamination, the Court drew attention to the importance of the role played in the matter by the Commission, which has a unique overview of developments in the nuclear power industry throughout the territory of the Community. Based on that consideration, the Court ruled that the Commission must be provided with general data relating to any plan for the disposal of radioactive waste before the competent authorities of the Member State concerned authorize such a disposal.[19]

In the *Nuclear Safety Convention* judgement, the Court went a step further by outlining the middle part of a judicial interpretation of the Euratom competencies in the field of nuclear safety that will be enriched even further by the most recent judgement in the case of the *Temelín Nuclear Power Plant* of 2009 (see Section 3.3.3).

In this decision, the ECJ added that it is inappropriate, in order to define the Community's competencies, 'to draw and artificial distinction between the protection of the health of the general public and the safety of sources of ionizing radiation'.[20] According to the Opinion delivered by the Advocate General Jacobs: 'In the light of current scientific knowledge, it is neither possible nor desirable to maintain artificial barriers between the disciplines of radiation protection and nuclear safety'.[21]

Thus, in a truly broad reading of the Treaty, the ECJ ruled that even though the Euratom Treaty does not grant the Community competence to authorise the construction or operation of nuclear installations, under Articles 30 to 32 of the Euratom the Community:

> possesses legislative competence to establish, for the purpose of health protection, an authorisation system which must be applied by the Member States. Such a legislative act constitutes a measure supplementing the basic standards referred to in article 30 of the EURATOM Treaty.[22]

Based on this landmark ruling, the existing basic safety standards, aiming mainly at the protection of the health of workers and of the general public against the dangers arising from ionizing radiation, can thus be 'supplemented', within the meaning of the EURATOM Treaty, with safety requirements governing the safe management of radioactive waste and spent fuel.

19 *Land de Sarre et al*, above n. 18, paras. 11-20.
20 *Commission v. Council of the European Union*, above n. 16, para. 82.
21 Opinion of AG Jacobs, delivered on 13 December 2001, ECR 2002 I-11225, para. 166.
22 *Commission v. Council of the European Union*, above n. 16, para. 89.

Following this case, there were no longer any *legal* obstacles (political obstacles still remain) for the Commission to start the process of adopting legislation on nuclear safety for installations and on nuclear waste.[23]

3.3.2 Commission v. UK (II) *and* (III)*: The 'Military Activity' Cases*

The applicability of the Euratom Treaty to military activities has been the object of discussion since the mid-1990s. In 1995, three individuals, living in French Polynesia, in the vicinity of the place where French authorities announced that they intended to carry out a series of nuclear tests, sought to annul the Commission's Decision of 23 October 1995, by which the Commission concluded that the tests in that area did not present a perceptible risk of significant exposure for workers or the general public and that the basic standards for health protection would be met. In the case before the European Court, both the Commission and Parliament argued that the Euratom Treaty applied both to civil and military experiments.[24] French Government submitted that the provisions of Chapter 3 of the Treaty do not apply to nuclear activities in the military sphere.[25] However, the Court decided not to give answers about the substantial side of the matter, thus leaving the issue unresolved and implicitly admitting that military activities *might fall within* the scope of application of the Euratom Treaty.

Furthermore, in the *Nuclear Convention* case, the Court stated in the broadest terms that Article 37 Euratom gives the Community competences:

23 See: Sousa Ferro M., 'Nuclear Law at the European Court in the 21st Century', above n. 1, p. 78, and Sousa Ferro M., 'The Future of the Regulation of Nuclear Energy in the EU', International Journal of Nuclear Law (vol. 2, no. 2), 2008, p. 21. The immediate consequence of this judgement was that the Commission decided to amend the Declaration annexed to its Decision ratifying the 1994 Convention on Nuclear Safety in accordance with the ECJ judgement, namely that Euratom has competence in the field of nuclear safety. See: Commission Decision 2004/491/Euratom of 29 April 2004, OJ L 172/7. A far-reaching analysis of the consequences of this judgement was conducted in a document prepared by the Council's legal service in October 2003 (see Council Doc. No. 13909/03). Unfortunately, the version of this document that was made public did not include all the relevant and substantial parts of the original document.

24 Order of the President of the Court of First Instance of 22 December 1995, *Marie-Thérèse Danielsson, Pierre Largenau and Edwin Haoa v. Commission of the European Communities* (T-219/95 R), ECR 1995 II-03051, para. 12. The Commission referred the applicability of the Treaty to Art. 34, which states that 'Any Member State in whose territories particularly dangerous experiments are to take place shall take additional health and safety measures, on which it shall first obtain the opinion of the Commission' and that 'The assent of the Commission shall be required where the effects of such experiments are liable to affect the territories of other Member States.' See: Consolidated Version of the Treaty Establishing the European Atomic Energy Agency, as it results from the amendments introduced by the Treaty of Lisbon, signed on 13 December 2007 and which entered into force on 1 December 2009, Luxembourg: Publications Office of the European Union, 2010.

25 See: Order of the President, *Marie-Thérèse Danielsson et al*, above n 24, para. 33.

as regards any plan for the disposal of radioactive waste in whatever form if the implementation of that plan is liable to result in the radioactive contamination of the water, soil or airspace of another Member State.[26]

The still open question of the applicability of the Euratom Treaty to military activities was finally resolved by the Court in mid-2005, with a Decision, twice confirmed since, that no provision of the Euratom Treaty applies to military activities. These Judgements (the so-called 'Military Activities' cases) resulted in great surprise and consternation because they showed a clear shift in the ECJ's case law in the legal basis of radiological protection, departing from the background of the *Nuclear Safety Convention* case, that is a ruling which unequivocally demonstrates the will of the ECJ to set itself as a strong supporter of the expansion of the transfer of powers from the Member States to the Community, even in a decisively sensible area such as that of nuclear safety.

The 'Military Activities' cases started with the judgement of 12 April 2005 in the case *Commission v. United Kingdom*[27] concerning the decommissioning of the reactor Jason, a nuclear reactor used for training and research associated to the Royal Navy's nuclear propulsion programme for submarines. By its application, the Commission sought from the Court a declaration that by failing to provide general data relating to the plan for the disposal of radioactive waste associated with the decommissioning of the Jason reactor, the United Kingdom failed to fulfil its obligation under Article 37 of the Euratom Treaty.[28] According to the Commission, Article 37 applies, in fact, to the disposal of radioactive waste from both civil and military installations, as it aims to prevent any risk of radioactive contamination of another Member State and because, since the protection of the general public against the dangers arising from ionizing radiation is an invisible objective, it must extend to all sources of danger, including those resulting from the decommissioning of military installations, such as the Jason reactor. The United Kingdom, supported by the French Republic (the other great nuclear power in the EU), replied that Article 37 cannot

26 *Commission v. Council of the European Union*, above n. 16, para. 103. See also: Judgement of the Court of 4 October 1991, *European Parliament v. Council of the European Communities* (C-70/88), ECR 1991 I-04529, para. 14, where the Court held that: 'the purpose of the articles referred to [Art. 30 Euratom treaty] is to ensure the consistent and effective protection of the health of the general public against the dangers arising from ionizing radiations, whatever their source and whatever the categories of persons exposed to such radiations'.

27 Judgement of the Court (Grand Chamber) of 12 April 2005, *Commission of the European Communities v. United Kingdom of Great Britain and Northern Ireland* (*Commission v. UK (II)*) (C-61/03), ECR 2005 I-02477.

28 Under this Art.: 'Each Member State shall provide the Commission with such general data relating to any plan for the disposal of radioactive waste in whatever forms will make it possible to determine whether the implementation of such plan is liable to result in the radioactive contamination of the water, soil or airspace of another Member State. The Commission shall deliver its opinion within six months, after consulting the group of experts referred to in Article 3'. See: Consolidated Version of the Treaty Establishing the European Atomic Energy Agency, above n. 24.

apply to the disposal of radioactive waste from military installations since, on the one hand, the Treaty itself covers only the civil uses of nuclear energy and, on the other hand, the provisions of Chapter 3 of the Euratom Treaty (regarding health and safety) cannot have a scope wider than that of the provisions in other Chapters of the same Treaty. Finally, the United Kingdom and the French Republic argued that if it was the intention of the signatories of the Treaty to ensure applicability to military activities, they would have included in the Treaty itself some derogating provisions specifically intended to safeguard the national defence interests of the Member States (such as Article 48 para. 4 of the EEC Treaty, now Article 45 of the TFEU; and Article 223 EEC Treaty, now Article 346 of the TFEU).

At the beginning of its judgement, the Court highlighted that the signatories of the Treaty:

> by referring in the preamble thereto to the advancement of the cause of peace, the applications of the nuclear industry contributing to the prosperity of their peoples and the peaceful development of atomic energy, intended to emphasise the non-military character of that Treaty and the supremacy of the aim of promoting the use of nuclear energy for peaceful purposes.[29]

Furthermore, the Court observed that both Articles 1 and 2 (which define the mission as well as the tasks entrusted to the Euratom Community respectively) confirm the fact that 'the objectives pursued by the Treaty are essentially civil and commercial'. Moreover, the Court stated that in the absence of an express provision excluding activities connected to defence from the scope of the Treaty 'it is necessary to have regard to other factors in order to determine whether the Treaty is intended also to govern, at least in certain spheres, the use of nuclear energy for military purposes'.[30]

Moving on from this legal background, the Court did not consider as a decisive factor that some provisions of the Euratom Treaty deal specifically with the military sector (namely Articles 24 to 28 and 84). Sharing the objections made by the United Kingdom and the French Republic, the Court observed that:

> the existence of those provisions may also be explained by the fact that the application of certain rules introduced by that Treaty, even if it relates only to civil activities, is nevertheless liable to have an impact on activities and interests within the field of the national defence of the Member States.[31]

29 *Commission v. UK (II)*, above n. 27, para. 26.
30 *Commission v. UK (II)*, above n. 27, para. 28.
31 *Commission v. UK (II)*, above n. 27, para. 32.

The Court didn't seem convinced by the little guidance provided for by the *travaux préparatoires*. According to the (rather simplistic view) of the Court on this matter, it is apparent, from the declarations mentioned in the *travaux préparatoires*, that the representatives of the States who took part in the negotiations 'held differing opinions on that issue and that they decided to leave it unresolved'.[32]

However, this observation is not totally convincing. As the Advocate General Geelhoed emphasized in his opinion, it should be remembered that the discussions took place in the context of international disarmament negotiations, as well as general speculation that the French Republic was considering joining the ranks of the nuclear military powers. The potential application of the future Euratom Treaty to nuclear defence was thus, concluded the AG, 'not at the time strictly necessary (...) [it was] an issue of certain political sensibility'.[33]

Apart from these considerations, the ruling of the Court is based on two juridical observations.

First of all, considering the substantial powers, which the Euratom Treaty gives to the Commission in order to intervene actively by means of legislation or in the form of an Opinion containing individual Decisions in various spheres of activity which are concerned with the use of nuclear energy, the ECJ highlighted that:

> it is clear that the application of such provisions to military installations, research programmes and other activities might be such as to compromise essential national defence interests of the Member States. Consequently (...) the absence in the Treaty of any derogation laying down the detailed rules according to which the Member States would be authorised to rely on and protect those essential interests leads to the conclusion that activities falling within the military sphere are outside the scope of that Treaty.[34]

Furthermore, the Court rejected the interpretation of Article 37 that the Commission put forward during the oral procedure, namely that Member States are obliged to communicate only information as well as general data concerning equipment or installations which are no longer assigned to military use and which the Member State concerned for that reason classified as 'waste'. However, an interpretation of Article 37 to the effect that the Member State concerned might decide both the time from which a military source of radioactive waste must be regarded as civil waste and the actual content of the data, which must be

32 *Commission v. UK (II)*, above n. 27, para. 29.

33 See: Opinion of the AG Geelhoed delivered on 2 December 2004, ECR 2005 I-02510. For more on this topic, see also: Andres-Ordax B., 'Radiological Protection and Military Activities: Recent European Case-Laws', Nuclear Inter Jura 2007 – Proceedings, Brussels, 2008, pp. 546 et seq.

34 *Commission v. UK (II)*, above n. 27, para. 36.

communicated to the Commission, would be in contradiction with the purpose of that provision. According to the Court, first, any late communication of the data would render nugatory the objective of prevention. Second, any partial communication of the relevant data would make it impossible to deliver an opinion with full knowledge of the facts.[35]

Though it showed some sympathy for a better protection of the population against nuclear radiation,[36] the Court did not share the Commission's opinion that Euratom Secondary Law applies to the decommissioning of a military nuclear reactor and ruled that Euratom Law generally does not apply to military installations as well as activities:

> the Commission has not demonstrated that the application of Article 37 EA to the decommissioning of the military installation in question is justified.[37]

Immediately after, the Court restated the same conclusion in general terms,[38] thus creating some confusion about the possibility that it has left the door open for further discussion of the matter or it has intended to definitely resolve the issue against the applicability of the Euratom Treaty to military activities.

In his Opinion, Advocate General Geelhoed, who strongly disagreed with the Court's conclusion, proposed an inclusive solution, based on a case by case approach more in line with the intention of the original drafters of the Treaty.[39]

The question was finally resolved less than one year later, when the Court upheld its view against the applicability of the Euratom Treaty to military activities with even a stronger language.

---

35 *Commission v. UK (II)*, above n. 27, para. 40. See also para. 41, where the Court observed that: '(...) an interpretation of Article 37 EA which allowed Member States such discretion as to the time for communicating data and its content would be a source of dispute and would undermine the effective application of that provision'.

36 *Commission v. UK (II)*, above n. 27, para. 44.

37 *Commission v. UK (II)*, above n. 27, para. 43.

38 *Commission v. UK (II)*, above n. 27, para. 44, 'It is necessary, however, to emphasise that the fact that the Treaty is not applicable to uses of nuclear energy for military purposes.'

39 His Opinion, see above n. 33, is mainly based on two considerations. On the one hand, he observed that: 'The exclusion of all military activities from the scope of Euratom would mean that the provisions of the Treaty dealing expressly with its application to defence would be redundant and serve no purpose'. On the other hand, considering that '[T]he health and safety provisions of the EURATOM Treaty are of vital importance and should be interpreted in a manner that ensures effective protection of public health', the AG at first observed that 'the obligation imposed by Article 37 on member States to provide the Commission with general data relating to any plan for the disposal of radioactive waste should, in principle, apply equally to the defence sector'. However, he concluded that 'Article 37 obligations should not (...) apply where, in a particular case, a Member State considers that its essential security interests may be harmed by supplying certain information required under this article.'

The case *Commission v. UK (III)* concerned the lack of implementation by the United Kingdom of a Directive in Gibraltar, specifically a failure to inform the public likely to be affected in the event of a radiological emergency about the local emergency plan.[40]

According to the Commission, the information which Member States are required, under Article 5 of the Directive, to give to the public about health protection measures to be adopted in the event of a radiological emergency 'is a civil defence matter and not a military matter'.[41] The Court dismissed this argument, stating that 'it is common ground that in the present case the source of the nuclear energy is of military origin'.[42] Therefore, since the two cases are similar, in that they concern activities within the military sphere, the Court relied on the dictum of *Commission v. UK (II)* and ruled, in the broadest and clearest way, that:

> It is very clear from that judgment that the use of nuclear energy for military purposes falls outside the scope of all the provisions of the EAEC Treaty, not just some of them.

With this judgement, the ECJ confirmed with an even stronger language that military activities are outside the scope of application of both the Euratom Treaty and its secondary legislation (including, inter alia, the implementation of radiological protection provisions concerning emergency response).[43]

These rulings of the ECJ, which were once more confirmed in an Order of 2011 in a case brought before the Court claiming compensation for damages arising from the 1968 Thule nuclear accident,[44] attracted great consternation from most of the commentators. The Court unequivocally asserted its intention to exclude military activities from the scope of application of the Euratom Treaty and any concerns about the safety of persons and the

40 Judgement of the Court (First Chamber) of 9 March 2006, *Commission of the European Communities v. United Kingdom of Great Britain and Northern Ireland* (*Commission v. UK (III)*) (C-65/04), ECR 2006 I-02239.

41 *Commission v. UK (III),* above n. 27, para. 21.

42 *Commission v. UK (III),* above n. 27, para. 23.

43 Even AG Geelhoed, in his Opinion, agreed with the Court, albeit with a lack of enthusiasm. Not only, he observes, 'it is incontrovertible that the source of nuclear energy at issue in the present case, a nuclear submarine forming part of the UK Royal Navy, is military in nature', but it is also clear from the statements in the judgement *Commission v. UK (II)* that 'the Court's conclusion at paragraph 44 of the judgment that "the Treaty is not applicable to uses of nuclear energy for military purposes" is categorical and absolute.' 'It follows', he concluded in a resigned spirit, 'that for as long as the Community has not made use of its competence under the EC Treaty to legislate in this sphere, a gap exists in the protection of the health of the general public. It is clear from the judgement's terms that the Court has accepted this consequence. For these reasons I am compelled to conclude that the Commission's action in the present case must fail.' See: Opinion of AG Geelhoed delivered on 1 December 2005, ECR 2006 I-02251, paras. 28-37.

44 See: Order of the Court (Fifth Chamber) of 12 January 2011, *Eriksen et al v. Commission* (C-205/10 P, C-217/10 P and C-222/10), ECR 2011 I-00001, para. 66.

environment in case of radioactive fallout from a military use of nuclear energy or ionizing radiation clearly was, quite surprisingly, set aside. The question is, therefore, how is it possible to explain such a sudden shift by the Court from its previous assertion (in the *Nuclear Safety Convention* case) that the purpose of the Treaty's rules on radiological protection 'was to ensure consistent an effective protection of the health of the general public against the dangers arising from ionising radiations, whatever their source'.[45]

Some commentators observed that, even if from a functional point of view, the Commission's teleological argumentation to apply Euratom's health and safety policy to marginal military nuclear activities, especially when European civilians are endangered, might be considered as a reasonable option. The Court probably wanted to prevent it for (not better explained) political reasons linked to the military sphere.[46] Another commentator, recollecting the passage of the *Commission v. UK (II)*, where the Court suggested that the relevant provisions of the TFEU could be indeed used to adopt the appropriate measures for the protection of the health of the public and the environment against the dangers related to the use of nuclear energy including military purposes, observed that such a view seems to suggest that the ECJ has possibly grown tired of the Euratom Treaty, a Treaty that Member States seem to have given up on.[47]

If we limit our analysis to the *Military Activities* cases, we probably might be inclined to support the latter theory, namely that the Court has grown tired of the Euratom Treaty, therefore not wanting to support a further expansion of the sovereign powers from the Member States to the Euratom Community. If we move on our analysis to other Judgments of the Court in the nuclear safety area, we then probably need to re-evaluate the theory of a 'political restraining' that inhibited the Court from carrying on its previous body of case law in favour of a broad interpretation of the Euratom Treaty in so far as military activities are concerned. What makes the difference between these two extremes is the *Temelín* judgement that the Court ruled about in 2009, a historical ruling that, in many ways, 'can be seen as an important step forward in consolidating and developing Euratom law'.[48]

3.3.3 *The* Temelín *Judgement: Who Ultimately Decides the Level of Nuclear Radiation Protection in the EU?*

In 2001, the Province of Upper Austria (acting as a private landowner, and not in its capacity as a public authority), considering that the radioactivity from the Temelín power

45 See: *Commission v. Council of the European Union,* above n. 16, para. 80.

46 See: Wolf S., 'Euratom before the Court: A Political Theory of Legal Non-Integration', European online Papers (EIoP), vol. 15.

47 See: Sousa Ferro M., above n. 1, p. 84.

48 See: Möstl M., 'Case C-115/08, *Land Oberösterreich v. CEZ,* Judgement of the Grand Chamber of 27 October 2008, in Common Market Law Review (47), 2010, p. 1230.

plant (located in the Czech Republic, at about 60 kilometres from the Austrian border) and the risks of contamination as well as malfunction were detrimental to the use of its agricultural college, located near the Czech border, brought action against the activities of the power plant before the *Landesgericht* (Regional Court) of Linz. Applicable Austrian Law indeed permits a landowner to have the nuisance emanating from his neighbour's installations prohibited. However, operators granted official authorisation issued by the Austrian authorities may avoid such action and the owner, in that specific case, can merely apply for compensation. What we have in this case is that the Temelín power plant was authorised only by the Czech authorities, and not by the Austrian. The Austrian Supreme Court (*Oberster Gerichtshof*) in a judgement of 4 April 2006 decided that authorisations granted by Austrian authorities had that effect. According to the *Oberster Gerichtshof*, the reason is that the relevant Article (para. 364 of the Austrian Civil Code) is based exclusively on the consideration of diverging national interests and there is no reason why Austrian Law should restrict the property rights of Austrian landowners in the interest of protecting a foreign economy as well as public interests in another country. According to the *Landesgericht* of Linz, however, that interpretation could be contrary to Community Law and thus it asked the ECJ whether the interpretation of the *Oberster Gerichtshof* could be compatible with certain provisions of the EC Treaty, in particular with the fundamental freedoms as guaranteed in Articles 10 (principle of loyal cooperation), 12 (principle of non-discrimination), 28 (prohibition of measures having equivalent effect), and 43 (prohibition of restrictions on the freedom of establishment of nationals of a Member State in the territory of another Member State).[49]

The Grand Chamber of the ECJ ruled against the Province of Upper Austria. First of all, the Court noted that, the present case being a dispute involving an industry activity consisting in the operation of a nuclear power plant, it falls within the scope of application of the Euratom Treaty (*lex specialis* principle).[50] This is important, because both the

49 See: Judgement of the Court (Grand Chamber) of 27 October 2009, *Land Oberösterreich v. CEZ* (C-115/08), ECR 2009 I-10265, paras. 38-54. The construction and operation of the Temelín nuclear power plant were authorised by the Czech authorities in 1985 and it began operating on a trial basis on 9 October 2000. Since 2003, it has operated at full capacity. On 29 November 2001, Austria and the Czech Republic adopted a document, known as 'The Conclusions of the Melk Process and Follow-Up' annexed to the final act of the Treaty concerning the accession of 10 new Member States, including the Czech Republic, signed in Athens on 16 April 2003 (OJ 2003 L 236, p. 17), in which both States declared that they would fulfil the series of bilateral obligations set out in those conclusions. Parallel with the Melk process, the safety of the Temelín nuclear power plant was evaluated by the Commission as well as the Council, and the results of that evaluation showed that the Temelín nuclear power plant, subject to the implementation of the proposed recommendations, showed a satisfactory level of nuclear safety. Since the accession of the Czech Republic to the EU, additional checks have been carried out pursuant to Art. 35 Euratom, in which Temelín has been found compliant with the prevailing legislation and a definitive declaration has been issued to that effect.

50 *Land Oberösterreich v. CEZ*, above n. 49, para. 83. This was done by recalling that, under Arts. 234 EC and 150 Euratom, the Court has 'identical jurisdiction for the purpose of interpreting the relevant provisions of the EC and EAEC Treaties'.

referring court and Advocate General Maduro directly relied on EC Law.[51] In particular, the Court referred to Article 12 of the EC Treaty, which prohibits any discrimination on grounds of nationality:

> is a specific expression of the general principle of equality, which itself is one of the fundamental principles of Community law. It must therefore be recognised that (...) the principle of prohibition of any discrimination on grounds of nationality (...) is a general principle which is also applicable under the EAEC Treaty.[52]

It was then quite simple for the Court to conclude that the difference in treatment introduced by Austrian Law, which worked to the detriment of installations, which received official authorisation in a Member State other than the Republic of Austria, 'in reality leads to the same outcome as a difference in treatment on grounds of nationality'.[53]

At this point, the Court, in order to demonstrate that the difference in treatment did come within the scope of application of the Euratom Treaty, once again elaborated on the competencies of the Community in the nuclear energy field. It stated that although the Euratom Treaty does not contain a Title relating to nuclear installations, Title II of the Treaty has a Chapter 3 entitled 'Health and Safety' which is intended to provide for the protection of public health in the nuclear sector. In particular, Articles 35 to 38 Euratom form, in the words of the ECJ, 'a coherent whole conferring on the Commission powers of some considerable scope on order to protect the population and the environment against the risks of nuclear contamination'.[54]

On the one hand, the ECJ based its conclusion on the fact that both the Euratom Community and its Member States are Parties to the 1994 Convention on Nuclear Safety (whose objective was, inter alia, to establish and maintain effective defences in nuclear installations against potential radiological hazards in order to protect human health as well as the environment). On the other hand, the Court reminded that, in the event of

51 See: Wolf S., above n. 1, p. 1644, who observes that: 'this demonstrates once again that EURATOM is hardly visible and its law is unknown to many legal experts'. See also: Möstl M., above n. 48, p. 1228, arguing that with its Decision to examine the case under the Euratom Treaty the Court avoided 'some weaknesses of the A.G's line of argument: reliance on a new and arguable extended scope of application of the fundamental freedoms; no clear statement on the relationship between EC and Euratom law in the present case'. The Author distinguishes the two approaches chosen by the A. G. and the Court in the following terms: 'The approaches of the Advocate General and of the Court represent two different styles of tackling those difficulties, an ambitious, innovative, but also somewhat uncertain and partly arguable one, as far as the legal basis and its exact consequences are concerned (A.G.), and a much more modest, but in the end also more reliable one (Court).'

52 *Land Oberösterreich v. CEZ*, above n. 49, para. 89.

53 *Land Oberösterreich v. CEZ*, above n. 49, para. 97.

54 *Land Oberösterreich v. CEZ*, above n. 49, para. 118.

malfunction of the protection system introduced by the Euratom Treaty, the Member States had a number of remedies at their disposal for obtaining the corrections necessary in the circumstances.[55] Among others, reference is made to the right to request the revision of the basic safety standards (Article 32 Euratom), the right to initiate infringement proceedings (Articles 142 and 38 Euratom in cases of urgency), and, finally, to the judicial review mechanisms concerning both the lawfulness of the measures taken by the Council or the Commission and cases where one of those institutions may even be brought before the Court immediately (Articles 145 to 149 Euratom).

In these circumstances, concluded the Court, Member States cannot exclude the justifiability of an authorisation granted regarding nuclear installations situated in other Member States by maintaining that such an exclusion is justified on grounds of protecting life, public health or the environment:

> Such an exclusion disregards completely the fact that the Community legislative framework (...) contributes precisely and essentially towards ensuring such protection.[56]

In contrast with the Opinion of Advocate General Maduro, the Court therefore concluded that once nuclear related activities meet the requirements of National Laws for protection against ionizing radiation provided for in Articles 30 to 33 Euratom, and once they passed the Community's scrutiny, they are no longer subject to discriminatory acts in other Member States.

However, by doing that the Court did not take into account that populations in Europe have different attitudes towards nuclear energy,[57] and therefore that it may be crucial for a Member State government to lay down stricter rules in order to meet the needs of its citizens. That is to say, that with its judging the ECJ deemed it necessary to answer the crucial question of who ultimately decides the level of nuclear radiation protection, but not taking into consideration that this is a political, rather than a juridical question.

55 *Land Oberösterreich v. CEZ*, above n. 49, para. 127-134.

56 *Land Oberösterreich v. CEZ*, above n. 49, para. 136.

57 See: European Commission, 'Attitudes towards Radioactive Waste', Special Eurobarometer 297Report, June 2008, p. 7, about the question 'Are you totally in favour, fairly in favour, fairly opposed or totally opposed to energy production by nuclear power stations?'. Only 14% of the Austrian population (the lowest mark beside Cyprus) answered 'Totally in favour', while 64% of the Czech population (the highest mark in all the EU 27) answered in that manner. See also: European Commission, 'Europeans and Nuclear Safety', Eurobarometer 324 Report, March 2010, p. 41, where 59% of the Czech population considered that the advantages of nuclear energy are greater that the risks it poses, in opposition to only 24% of the Austrian population sharing this view.

3.4 Conclusions

There are circumstances that seem to introduce some possible, significant elements of critique about the opportunity to assert the effectiveness of the Court's role in the area of nuclear safety.

First, the EU chose mainly to use Directives to legislate on the subject of waste, thus allowing Member States some room to manoeuvre. Moreover, the Directives in the nuclear field set up a very flexible framework for the Member States, stating that 'national circumstances' will be taken into account when Member States develop their national framework.[58] Of course, these Directives are Hard Law, in the sense that they are legally binding. Non-complying States, as we have seen, are exposed to sanctions and litigation in the ECJ, but these Directives also include 'soft' elements, and therefore both the Commission as well as the ECJ need transparency and cooperation from States if they want their actions to be really effective.

Second, Court cases take a long time to conclude. Thirteen years, for example, elapsed between the initial complaint to the Commission in 1987 about uncontrolled waste disposal in the Kouroupitos river in Crete and the historic Court of Justice ruling that Greece must pay a penalty of 20,000 euro a day until it complied fully with the relevant EU Directives.[59] How is it possible to conciliate the role of the Court to effectively enforce the obligations laid down in the EU Directives with such long procedural times ? Which sanctions are the most effective and could tempt Member States to fulfil their obligations? Is a fine an effective, final sanction, or might it simply be a further cost to a country?

Notwithstanding all the difficulties and the relatively low number of decisions, it is unquestionable that the ECJ has played a decisive role in shaping EU Nuclear Law, not hesitating to make decisions that can influence in a strong way the future of the European policies in an area which is of crucial importance for the well-being of national citizens.

In the *Nuclear Safety Convention* case, the ECJ was guided by Euratom. Because of the dramatic renitence of the Commission to legislate in the nuclear sector, the Court adopted

---

58 See: Södersten A., above n. 4, p. 8. These Directives have therefore been described as 'a framework for cooperation which includes general and open-ended guidelines rather than rigid rules and straightforward safety standards'.

59 See: Judgement of the Court of 4 July 2000, *Commission of the European Communities v. Hellenic Republic* (C-387/97), ECR 2000 I-05047. This case concerned the operation of an illegal waste dump at a deep gorge at Kouroupitos in the region of Chania. The site was used to illegally dispose of domestic waste, and, for a certain period, quantities of hazardous waste (for example waste oils and batteries) as well as a range of commercial and industrial waste. Since January 1988, the Commission had been investigating the matter. On 7 April 1992, the ECJ ruled against Greece for the first time. See: Judgement of the Court of 7 April 1992, *Commission of the European Communities v. Hellenic Republic* (C-45/91), ERC 1992 I-02509. However, Greece failed to comply with this judgement and the Commission brought the matter before the Court for a second time, using for the first time the power under Art. 228 of the EC Treaty to fine a Member State for non-compliance with a previous Court judgement for breach of EU waste legislation.

a decision whose effect was to remove, by broadly interpreting the powers granted to the Community by Chapter 3 of Title II of the Euratom Treaty, any legal obstacles for the Commission to start the process of adopting legislation on nuclear safety for installations and on nuclear waste.

As to the identification of the real limit of the legislation powers under the Euratom Treaty, the Court was much more uncertain. First, in the very same *Nuclear Safety Convention* case it adopted quite a broad approach, stating that Chapter 3 of Title II of the Euratom Treaty 'form a coherent whole conferring upon the Commission *powers of considerable scope* in order to protect the population and the environment against the risks of nuclear contamination' (italics added).[60] Some years later, however, the Court ruled that Euratom Law generally does not apply to military installations and activities, thus seeming to conclude that ionizing radiation is dangerous if it originates from civil sources, but not if it arises from military installations. It is as if the Court ruled that the protection of national security interests is more relevant than ensuring the protection for the health of populations and the environment. With the 'Military Activities' cases, the Court sided with the Member States, as much as it weakened the unity of the EU legal order. The passage of *Commission v. UK (II)* where the Court made the idealistic suggestion that the relevant provisions of the TFEU could be used to adopt the appropriate measures for the protection of the health of the public and the environment against the dangers related to the use of nuclear energy, doesn't seem satisfactory either given the scarce will of the Member States to lose sovereign powers in this sector.[61]

With the *Temelín* judgement the Court did something different. On the one hand, as we have seen, this judgement reinforced the Euratom competencies in many ways, on the other hand, as many commentators have pointed out, with this judgement the Court strongly weakened the will of the population (which in Austria, as we have seen, is largely opposed to the use of nuclear energy), and discouraged both the population and NGOs in neighbouring States from taking judicial action against those Member States which decide to start the construction as well as operation of nuclear power plants with cross-border effects.[62]

Keeping this perspective in mind, the recent case law of the Court, as established especially in the *Temelín* judgements, would seem to confirm that, at least as far as the ECJ

60 See: *Land de Sarre et al v. Ministre de l'Industrie*, above n. 18, para. 11; later confirmed in *Commission v. Council of the European Union*, above n. 16, para. 89.

61 For more on this, see Chapter 2 by Sousa Ferro in this volume.

62 See: Schärf W.-G., 'The Temelin Judgement of the European Court of Justice', Nuclear Law Bulletin No. 85 (Vol. 1), 2010, p. 85. According to the Author, any claims should take into account not only the relevant provisions of the Euratom Treaty, but also the results of the checks carried out by the Commission in accordance with Chapter 3 EURATOM and National Laws in light of the Euratom Treaty.

is involved, the Aarhus approach[63] is far from having achieved the desired results in matters covered by the Euratom Treaty. On the other hand, this body of case law perfectly reflects the recent trend of the ECJ rulings each time national security interests of Member States are mixed up with the role of the EU executive and legislative actors. In such a sensitive area, as for example is the case with counterterrorism policy (and one can easily remember the *Kadi* case), or more precisely the case of nuclear safety, a delicate balance is required between, on the one hand, the protection of health and the fundamental rights of a democratic society as well as, on the other hand, the interests of the national and EU authorities, namely national security interests. Given the narrow power of intervention reserved to the European Parliament (which, in the case of the Euratom Treaty, is practically equal to nothing at all), the question is therefore: What is the role of the Court in finding this balance, and which decisions should be reserved for the executive?

The fundamental character of this question is much more emphasized in the case of the Euratom Treaty, where Member States still possess a relevant margin of discretion. It is therefore under this specific perspective that we might appreciate the decisive role that the ECJ has played in the last few years in shaping EU Nuclear Law: resisting the pressure of (at least some) very resentful Member States, replacing the inactivity of the EU executive authorities, and advocating (even if sometimes in a confusing manner) for the unity as well as internal coherence of the EU legal order.

---

63 See: Convention on Access to Information, Public Participation in Decision-Making and Access to Justice in Environmental Matters, Aarhus, 25 June 1998, entered into force 30 October 2011, United Nations, Treaty Series, vol. 2161, p. 447 (see also: <www.unece.org/fileadmin/DAM/env/pp/documents/cep43e.pdf>). The EC is part of the Convention since May 2005, see: Council Decision 2005/370/EC of 17 February 2005 on the conclusion, on behalf of the European Community, of the Convention on Access to Information, Public Participation in Decision-Making and Access to Justice in Environmental Matters, OJ L 124, pp. 1-3. In 2003, two Directives concerning the first and second 'pillars' of the Aarhus Convention were adopted: Directive 2003/4/EC of the European Parliament and of the Council of January 2003 on public access to environmental information and repealing Council Directive 90/313/EEC, OJ L 41, 14 February 2003, pp. 26-32, and Directive 2003/35/EC of the European Parliament and of the Council of 26 May 2003 providing for public participation in respect of the drawing up of certain plans and programmes relating to the environment and amending with regard to public participation and access to justice Council Directives 85/337/EEC and 96/91/EC, OJ L 156, 25 June 2003, pp. 17-25. Regarding the access to justice, on 24 October 2003 the Commission presented a proposal for a Directive of the European Parliament and of the Council on access to justice in environmental matters (COM(2003)624). For more on the Aarhus legislation in the EU, see: <http://ec.europa.eu/environment/aarhus/legislation.htm>. As to the obligations set forth in the Aarhus Convention regarding the access to justice, Art. 9 states that all persons who feel that their rights to access to information have been impaired (request for information ignored, wrongfully refused, inadequately answered) must have access, in the appropriate circumstances, to a review procedure under national legislation. Access to justice is also ensured in the event of the Convention's participation procedure being infringed. Access to justice is also allowed for the settlement of disputes relating to acts or omissions by private persons and public authorities, which contravene provisions of National Law relating to the environment.

4 Radioactive Waste Transport under Euratom and Other EU Law

Leonardo Massai

4.1 Introduction to the Transport of Radioactive Waste

Radioactive material is produced by the nuclear fuel cycle, but also used in several sectors such as medicine, agriculture, research, manufacturing, non-destructive testing and the exploration of minerals.

The transport of radioactive waste (RW) covers about twenty million different sizes of consignments of radioactive materials. These are routinely transported globally on public roads, railways and ships. The RW transported usually contains low levels of radioactivity. The shipment of RW is usually managed by specialized transport agencies that are using robust and secure containers. The transport of RW by sea is generally based on *ad hoc* ships built for that purpose. Since 1971, more than 20,000 shipments of used fuel and high-level waste (over 80,000 tonnes) have been transported over many million kilometres.[1] Specific legislation has been adopted at both the international and the European level to regulate the transport of radioactive waste. While EU legislation on the transport of RW is the subject of this chapter, at the international level Regulations are adopted by the International Atomic Energy Agency (IAEA) and consequently applied at the national as well as regional level.

The transport of RW is an integral part of the nuclear fuel cycle, which covers about 430 nuclear power reactors in operation in 32 countries. Within the fuel cycle services specialised facilities are being developed in various locations and thus the need to transport radioactive materials between the various facilities. Nuclear fuel cycle materials are usually transported in solid form.

1 World Nuclear Association website, August 2014, <www.world-nuclear.org/info/Nuclear-Fuel-Cycle/Transport/Transport-of-Radioactive-Materials/>.

Table 4.1 Radioactive material transport activities

From:	To:	Material:	Notes:
Mining	Milling	Ore	Rare: usually on the same site
Milling	Conversion	Uranium oxide concentrate ('Yellowcake')	Usually 200-litre drums in standard 6m transport container
Conversion	Enrichment	Uranium hexafluoride (UF_6)	Special UF_6 containers
Enrichment	Fuel fabrication	Enriched UF_6	Special UF_6 containers
Fuel fabrication	Power generation	Fresh (unused) fuel	
Power generation	Used fuel storage	used fuel	After on-site storage, large Type B casks
Used fuel storage	Disposal*	used fuel	Large Type B casks
Used fuel storage	Reprocessing	used fuel	
Reprocessing	Conversion	Uranium oxide	Called reprocessed uranium (RepU)
Reprocessing	Fuel fabrication	Plutonium oxide	
Reprocessing	Disposal*	Fission products	Vitrified (incorporated into glass)
All facilities	Storage/disposal	Waste materials	Sometimes on the same site

Source: World Nuclear Association website.

Radioactive waste can be classified in accordance with the following:
- Low-level waste (LLW) containing enough radioactive material to require action for the protection of people, but not so much that it needs shielding in handling or storage;
- Intermediate-level waste (ILW) requires shielding. If it has more than 4,000 Bq/g of long-lived (over 30 years half-life) alpha emitters it is categorised as 'long-lived' and requires more sophisticated handling as well as disposal ('Type A' packages);
- High-level waste (HLW) sufficiently radioactive to require both shielding as well as cooling, generates >2 kW/m^3 of heat and has a high level of long-lived alpha-emitting isotopes robust and very secure casks known as 'Type B' packages).

For each classification of RW a different type of packaging is required.

All EU Member States produce radioactive waste, which is generated by, inter alia, electricity production in nuclear power plants and radioisotope applications in medicine, industry, agriculture, research as well as education.

Figure 4.1 Electricity generated by nuclear energy in the EU

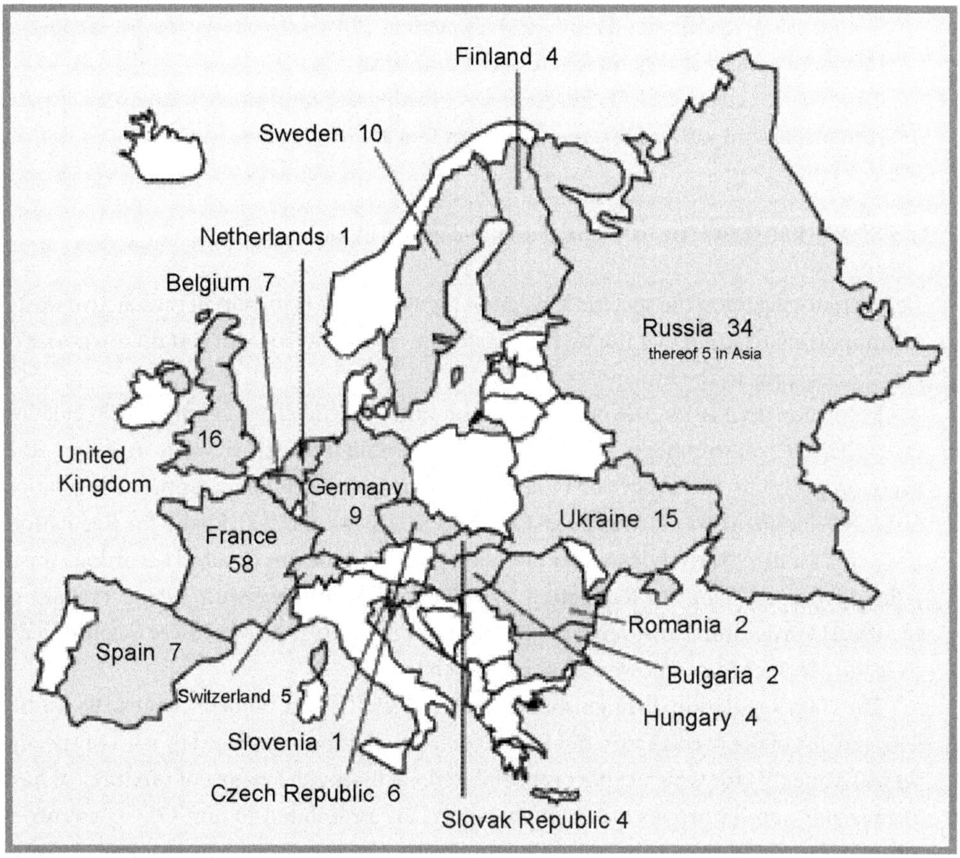

Source: European Nuclear Society, 2015

Each day thousands of radioactive material packages are transported within, between, into as well as out of EU Member States on land, water and by air. Of the dangerous goods transports in Europe 1% is RW. At the moment of writing, fourteen Member States have nuclear power reactors in operation, and a further two Member States have nuclear power reactors which are being decommissioned. Most MS have research reactors.

EU legislation covering the transport of radioactive waste material comprises the following instruments:

- Directive 92/3/Euratom;
- Council Directive 2006/117/Euratom on the supervision and control of shipments of radioactive waste and spent fuel;

- Commission Decision of 5 March 2008 establishing the standard document for the supervision and control of shipments of radioactive waste and spent fuel;
- Commission Recommendation of 4 December 2008 on criteria for the export of radioactive waste and spent fuel to third countries;
- Council Directive 2011/70/Euratom establishing a Community framework for the responsible and safe management of spent fuel and radioactive waste.

4.2 EU Legislation on the Transport of Waste

Before touching upon the specific legislation regulating the transport of radioactive waste, it is important to stress that not all EU legislation on the transport of certain categories of waste does apply to the transport of RW.

Regulation (EC) 1013/2006 on shipments of waste,[2] amended by Commission Regulation (EU) 135/2012 on shipments of waste to include certain unclassified waste in Annex III B thereto,[3] prohibits the export of hazardous waste to non-OECD countries. Regulation 1013/2006 implements the 1989 Basel Convention (Decision II/12). Under the Regulation, if waste is for disposal only, Member States may prohibit the import of all hazardous waste falling within the scope of Regulation 1013/2006 in accordance with Article 11(1)(e) of the Basel Convention. However, if the import is for recovery purposes, there is controversy over the Member States' powers to restrict such imports.

The 2006 Regulation distinguishes 'green' (Annex III) and 'amber' (Annex IV) waste.[4] The 'red' list of the predecessor, the 1993 Regulation, has been replaced by a list of specific waste (Annex V) for which export is prohibited, including the export of mixtures of hazardous and non-hazardous waste. Such exports are prohibited to non-OECD countries whether for disposal or recovery. The composition of the above lists in comparison to those of the Basel Convention and other relevant international instruments is a topic for further research. The distinction between intra-EU transport and transport outside the EU of the 1993 Regulation is maintained in the new instrument.

However, the Basel Convention does not apply to shipments of spent fuel or radioactive waste to a third country and neither does Regulation 1013/2006 implementing that Convention.[5] This is specified under Article 1.3.c of the 2006 Regulation, 'shipments of radioactive waste as defined in Article 2 of Council Directive 92/3/Euratom of 3 February

2 Regulation (EC) No. 1013/2006 of the European Parliament and of the Council of 14 June 2006 on shipments of waste, Official Journal of 12 July 2006, L 190/1, Brussels.

3 Commission Regulation (EU) No. 135/2012 of 16 February 2012 amending Regulation (EC) No. 1013/2006 of the European Parliament and of the Council on shipments of waste to include certain unclassified waste in Annex III B thereto, Official Journal of 17 February 2012, L 46/30, Brussels.

4 For green list waste, a general information requirement is stipulated, for amber waste a PIC procedure.

5 Cf. Art. 1(3)(c) of the Basel Regulation.

1992 on the supervision and control of shipments of radioactive waste between Member States and into and out of the Community'.

By the same token, Directive 2008/98/EC on waste (Waste Framework Directive) and repealing certain Directives[6] and Directive 2008/68/EC on the inland transport of dangerous goods[7] are also not applicable to the transport of radioactive waste (to be precise, the former may have a very small impact on the transport of radioactive waste). Directive 2008/98/EC regulates the protection of the environment as well as human health 'by preventing or reducing the adverse impacts of the generation and management of waste and by reducing overall impacts of resource use and improving the efficiency of such use'. Directive 2008/98/EC explicitly excludes radioactive waste from its scope of application as indicated in Article 2(1)d. Directive 2008/68/EC is about the 'transport of dangerous goods by road, by rail or by inland waterway within or between Member States, including the activities of loading and unloading, the transfer to or from another mode of transport and the stops necessitated by the circumstances of the transport' (Article 1). Radioactive waste is not specifically mentioned apart from the Annex dealing with transport by road and including some derogations for Member States for the transport of dangerous goods within their territory on the basis of Article 6(2) of the Directive. The transport of radioactive materials is in some cases mentioned explicitly in the Annex.

4.3 Directive 92/3/Euratom

For the transport of radioactive waste, the substantive provisions of the applicable IAEA and IMO Conventions were brought into special EU legislation by way of Directive 92/3/Euratom,[8] replaced by Directive 2006/117/Euratom.[9] In particular, this was the case for the 1997 IAEA Joint Convention on the Safety of Spent Fuel Management and on the Safety of Radioactive Waste Management.[10]

The legal basis for action of the European Commission in the area of transport of radioactive material lays in both the old European Community Treaty and the Euratom Treaty, respectively Title V on the common transport policy as well as Articles 31 and 32. In 1982, at the request of the European Parliament, the Commission established a Standing Working Group (SWG) composed of national experts with some competence on the

---

6 Directive 2008/98/EC of the European Parliament and of the Council of 19 November 2008 on, Official Journal of 22 November 2008, L 312/3, Brussels.

7 Directive 2008/68/EC of the European Parliament and of the Council of 24 September 2008 on the inland transport of dangerous goods, Official Journal of 30 September 2008, L 260/13, Brussels.

8 Council Directive 92/3/Euratom of 3 February 1992 regarding the supervision and control of shipments of radioactive waste between Member States into and out of the Community.

9 OJ 2006, L 337/21.

10 The Convention entered into force on 18 June 2001. Since 2005, Euratom has been Party to the Convention (see Decision 2005/84/Euratom, OJ 2005, L 30/10.

transport of radioactive material. Among other things, the SWG can make proposals for the Commission's action in the field of radioactive waste material and produce periodical reports.

Council Directive 92/3/Euratom of 3 February 1992 on the supervision and control of shipments of radioactive waste between Member States and into and out of the Community established a system of prior authorization for the shipment of waste and radioactive material. At the moment of writing, Directive 92/3 has been repealed by Directive 2006/117/Euratom which will be discussed below.

Directive 92/3 regulates the shipments of radioactive waste between Member States in and out of the Community whenever the quantities and concentration of radioactive waste exceed the levels laid down in Article 4(a) and (b) of Directive 80/836. Directive 92/3 defines '"radioactive waste" as any material which contains or is contaminated by radionuclides and for which no use is foreseen', and '"shipment" as transport operations from the place of origin to the place of destination, including loading and unloading, of radioactive waste'. In accordance with Article 3, international agreements on the transport of radioactive material are transposed into EU Law. Directive 92/3/Euratom introduces a compulsory and common system of notification as well as standard control.

In accordance with Article 4 of Directive 92/3, before carrying out a shipment of such waste or to arrange for such a shipment to be carried out the holder of radioactive waste must submit an application for authorization to the competent authorities of the country of origin. The competent authorities of the country of origin shall send such applications for approval to the competent authorities of the country of destination and of the country or countries of transit, if any. The authorization shall be valid for a period of not more than three years.

Council Regulation (Euratom) No. 1493/93 of 8 June 1993 on shipments of radioactive substances between Member States[11] applies to shipments, between Member States, of sealed sources and other relevant sources, whenever the quantities as well as concentrations exceed the levels laid down in Article 3.2(a) and (b) of Directive 96/29/Euratom. The Regulation's 1493/93 main points are:

- 'Sealed source': a source of ionising radiation consisting of radioactive substances firmly incorporated in solid and effectively inactive materials, or sealed in an inactive container of sufficient strength to prevent, under normal conditions of use, any dispersion of radioactive substances;
- A holder of sealed sources who wishes to carry out a shipment of certain substances provided for by the Directive must obtain a prior written declaration by the consignee of these substances, using the standard documents;

11 Council Regulation (Euratom) No. 1493/93 of 8 June 1993 on shipments of radioactive substances between Member States, Official Journal of 19 June 1993, No. L 148/1.

- The above declaration must be sent by the consignee to the competent authority of the Member State to which the shipment is to be made, valid for a period of not more than three years from the date of stamping by the competent authority.

4.4 COUNCIL DIRECTIVE 2006/117/EURATOM

Radioactive waste and in particular the shipment of radioactive waste are not mentioned in the Annexes to Regulation (EC) No. 1013/2006 of the European Parliament and of the Council of 14 June 2006 on shipments of waste that specifies under which conditions waste can be shipped between countries. The shipment of radioactive waste is regulated mainly by Directive 2006/117/Euratom.[12] The 2006 Directive has been modelled on the provisions and requirements for hazardous waste established under the Basel Convention such as notification as well as permission requirements for each transport. Decisions adopted within the OECD (the NEA), the IAEA and the IMO have been considered while framing the 2006 Directive.

The 2006 Directive was adopted mainly because of the clear need to amend Directive 92/3/Euratom with the view to:
- clarify and add concepts as well as definitions;
- address situations that had been omitted in the past and
- simplify the existing procedure of the shipment of radioactive waste between MS.

It applies to trans-boundary shipments of radioactive waste and spent nuclear fuel. Spent fuel is the nuclear fuel that has been irradiated and permanently moved from a reactor's core. Spent fuel may either be considered as a usable resource to be reprocessed or be destined for disposal if regarded as radioactive waste.

The 2006 Directive establishes a system of prior authorization for such shipments in Europe. The Member States concerned have to be informed, a system of periodical reporting has been introduced, and the system of prior authorization has been modified. The Member State of destination can refuse to accept the shipment – justification and export to the African, Caribbean, and Pacific Group of States (ACP) is prohibited. The holder is required to submit a duly completed application to the competent authorities of the Member State of origin and the shipment cannot be made without prior approval. Finally, the competent authorities may add conditions to the shipment.

In particular, the 2006 Directive:
- requires operators to notify national authorities about shipments of radioactive materials which depart from, go through, or end up in the EU;

12 OJ 2006, L 337/21.

- allows EU countries to ship spent fuel to each other for reprocessing and organise the return of the resulting radioactive materials;
- allows EU countries to send shipments of radioactive materials that do not comply with the Directive back to their country of origin;
- prohibits the export of radioactive waste to African, Caribbean or Pacific countries, to Antarctica, or to any country which does not have the resources to safely manage it.

The 2006 Directive maintains the distinction between intra-EU transport and transport outside the EU (as in the Regulations on the Transport of Hazardous Waste). Other than under the old Directive 92/3 Euratom, shipment of spent nuclear fuel for reprocessing is now also included in the 2006 Directive (under Article 1(2)). In the event that the shipment is contrary to international, European or national obligations, Member States of transit may refuse to give their consent (Article 9(3)(a)). Member States of destination may also invoke legislation on the management of radioactive waste or spent fuel (Article 9(3)(b)). In any case, conditions attached to the consent must be non-discriminatory, i.e., they may not be more stringent than requirements for similar shipments within that Member State.

A Standing Working Group (SWG) of national experts on the transport of radioactive materials was set up to assist the European Commission to develop new Regulations on the transport of radioactive materials. The main task and role of this Group is to exchange information on the application of Regulations concerning the transport of radioactive materials worldwide with a view to prepare proposals to the Commission. The most recent report from the SWG (fifth report 2006) recommended the implementation of a new radioactive waste transport safety programme covering six main areas:

- supporting the international review as well as revision of radioactive materials transport Regulations and safety guides;
- strengthening safety and security in the transport of radioactive waste regarding the latest scientific and technological developments;
- furthering the development of emergency preparedness and response to prevent illicit trafficking in radioactive materials;
- assisting newer EU countries in the development and implementation of national regulatory infrastructure;
- promoting transparency by providing information to the public and the media;
- reducing the refusal of safe shipments of radioactive materials and removing barriers to competition.

The European Commission adopted the following two provisions in this area:
- Commission Decision 2008/312/Euratom of 5 March 2008 establishing the standard document for the supervision and control of shipments of radioactive waste and spent fuel referred to in Council Directive 2006/117/Euratom;[13]
- Commission Recommendation of 4 December 2008 on criteria for the export of radioactive waste and spent fuel to third countries.[14]

4.5 DIRECTIVE 2011/70/EURATOM ('WASTE DIRECTIVE')

Apart from the 2006 Shipments Directive, Directive 2011/70 /Euratom on the management of spent fuel and radioactive waste also applies to the transport of such waste, at least partially. The scope of Directive 2011/70/Euratom extends to all stages of spent fuel and radioactive waste management.[15] More specifically, Article 4.2 says:

> (...)
> 2. Where radioactive waste or spent fuel is shipped for processing or reprocessing to a Member State or a third country, the ultimate responsibility for the safe and responsible disposal of those materials, including any waste as a by-product, shall remain with the Member State or third country from which the radioactive material was shipped.

In accordance with Article 16(2) of the 2006 Shipments Directive, the Member State where the radioactive waste is generated shall dispose of its radioactive waste, unless it has concluded a shipment agreement with another Member State or with a third country. That Member State is required to inform the Commission prior to a shipment to a third country of the content of such a shipment agreement, unless the shipment is directed to another Member State. The shipment agreement must satisfy specific safety and other criteria, stipulated in the 2011 Directive.[16]

---

13 Commission Decision of 5 March 2008 establishing the standard document for the supervision and control of shipments of radioactive waste and spent fuel referred to in Council Directive 2006/117/Euratom, 2008/312/Euratom, Official Journal of the EU, L 107, 17 April 2008, pp. 32-59.

14 Commission Recommendation of 4 December 2008 on criteria for the export of radioactive waste and spent fuel to third countries (notified under Doc. No. C(2008)7570), Official Journal of the EU, L 338, 17 December 2008, pp. 69-71.

15 Art. 2 states: '1. This Directive shall apply to all stages of: (a) spent fuel management when the spent fuel results from civilian activities; (b) radioactive waste management, from generation to disposal, when the radioactive waste results from civilian activities. 2. This Directive shall not apply to waste from extractive industries which may be radioactive and which falls within the scope of Directive 2006/21/EC.' (below n. 17).

16 Art. 4.4.a-c of the 2011 Directive. The agreement should be in accordance with Art. 16(2) of the 2006 Shipments Directive. Perhaps these requirements can avoid, in the case of radioactive waste, a disaster like

Directive 2011/70/Euratom introduced additional binding conditions in case of shipment of radioactive waste. As a general principle, Article 4(4) of that Directive indicates that radioactive waste must be disposed of in the Member State in which it was generated.[17] An exception to this general principle is that prior to a shipment to a third country, the Member State exporting the radioactive waste shall inform the Commission of the content of any such shipment and take reasonable measures to preserve the environment on the basis of a set of criteria established by the European Commission in accordance with Article 16(2) of Directive 2006/117/Euratom. Member States have the right to accept the spent fuel or waste for processing or reprocessing coming from another country and send it back to its country of origin. The ultimate responsibility for the safe and in charge of disposal of those materials remains with the Member State or third country from which the radioactive material was shipped.

4.6 Proposal for a Council Regulation Establishing a Community System for Registration of Carriers of Radioactive Materials

On 28 September 2012, the European Commission published a proposal for a Council Regulation establishing a Community system for registration of carriers of radioactive materials.[18] The main objective of the proposal is the introduction of a single registration system for conducting the transport of radioactive materials in the Member States. The single registration system will therefore replace the different national reporting as well as authorization procedures required under Articles 3 and 4 of Directive 96/29/Euratom and simplify the different procedures and administrative burdens. The Regulation will establish a European System for Carrier Registration. The carriers should function through a central web interface. Applicants will be required to fulfil some basic safety standards in order to obtain the registration by the respective national competent authority. In particular, the proposed Directive will:

- facilitate Member States' compliance with the basic safety standards for the protection of the health of workers and the general public against the dangers arising from ionising radiation;

the 2006 *Probo Koala* dumping of toxic waste in Côte d'Ivoire (see, for an overview of this toxic waste incident, which also led to numerous court cases, e.g., <www.marineinsight.com/misc/marine-safety/the-probo-koala-toxic-ship-incident-consequences>).

17 'Radioactive waste shall be disposed of in the Member State in which it was generated, unless at the time of shipment an agreement, taking into account the criteria established by the Commission in accordance with Article 16(2) of Directive 2006/117/Euratom, has entered into force between the Member State concerned and another Member State or a third country to use a disposal facility in one of them', Art. 4(4), Directive 2011/70/Euratom.

18 Proposal for a Council Regulation establishing a Community system for registration of carriers of radioactive materials, COM(2012)561, Brussels, 28 September 2012.

- apply to any carrier transporting radioactive materials within the Community, from third countries into the Community and from the Community into third countries;
- issue valid registrations for carriers of radioactive materials in accordance with its Article 5.

4.7 REPORT FROM THE COMMISSION ON THE IMPLEMENTATION BY THE MEMBER STATES OF COUNCIL DIRECTIVE 2006/117/EURATOM

On 25 April 2013, the European Commission published a report on the implementation of Directive 2006/117/Euratom (Shipments Directive)[19] covering the period 2008 to 2011. The report confirmed that the transposition of the Directive has been completed since the end of 2010 and provided for a feedback on the implementation of the general provisions of Directive 2006/117/Euratom as well as an overview of the Member States' information on the implementation of the Directive.

The Shipments Directive requires Member States to use a standard document for all shipments within the scope of the Directive. The Member States have reported some difficulties in using this document. The shipments of radioactive waste and spent fuel are subject to the standard document that includes the following information:

- acknowledgement of receipt of application;
- authorisation or refusal of shipment;
- description of consignment/list of packages;
- acknowledgement of receipt of shipment.

Under the Shipments Directive, competent authorities are those empowered to implement the supervision and control of shipments of radioactive waste or spent fuel. Article 19 of the Shipments Directive requires the establishment of a system of transmission of documents and information. In this respect, the Commission established a website containing all relevant information related to the Shipments Directive that is updated based on the information transmitted to the Commission by each Member State.

The Advisory Committee is composed of representatives from the Member States and has considered the following items until now:

- the establishment and use of the standard document;
- the Commission Recommendation establishing criteria for the export of radioactive waste and spent fuel to third countries;

19 Report from the Commission to the European Parliament, the Council and the European Economic and Social Committee on the implementation by the Member States of Council Directive 2006/117 EURATOM on the supervision and control of shipments of radioactive waste and spent fuels, Brussels, 25 April 2013, COM(2013)240 final.

- the Commission Recommendation for a secure and effective system of transmission of documents.

So far, the Advisory Committee has turned out to be very useful to ensure communication and cooperation among the Member States as well as the Commission.

Under the Shipments Directive Member States were required to report to the Commission for the first time by the end of 2011 and then every three years. The Member States, in the implementation of the Shipments Directive in the period considered, encountered no major problems. Some concerns were reported on the following matters:

- no harmonisation within the EU;
- transboundary shipments of some waste containing Naturally Occurring Radioactive Materials (NORM waste) not arising from authorized practices.

The shipments of radioactive waste and spent nuclear fuel between Member States are subject to the Prior Informed Consent Procedure of the competent authorities of the Member States involved. In the period considered, 14 Member States authorized shipments (2008-2011) and 161 authorizations were delivered (74% shipments of waste and 26% shipments of spent nuclear fuels).

Concerning the exports out of the European Union, the Commission established criteria facilitating the Member States to evaluate whether requirements for exports are met. In total, 9 Member States reported a total of 28 authorizations for export outside the EU to Russia, China, Switzerland, Japan and USA.

5 The Complexity of Radioactive Waste Management and Human Rights: The Case of the 'Quirra' Test/Training Range

Mario Odoni

5.1 Introduction

Quirra, or more exactly the 'Salto di Quirra', is an area in the south-east of the Region of Sardinia (Italy), between the Ogliastra and Cagliari provinces. Here the biggest Italian 'Inter-Forces' test/training range is based, the 'Poligono Sperimentale di Addestramento Interforze del Salto di Quirra'. This military range consists of a land area (Municipality of Perdasdefogu, Ogliastra), a quadrilaterally shaped upland of about 12,000 ha and a sea area (Capo S. Lorenzo, Sarrabus-Gerrei), which includes a coastal strip of 1,100 ha.

The Italian Air Force Staff, who needed a sparsely populated wide area to test new weapon systems, especially rockets and missiles, created the range in 1956. For these reasons, the 'Salto di Quirra' was chosen.[1] In 1959, the test range was converted into an 'Inter-Forces' range, for training of all military personnel, also from the Army and Navy. Even several national and foreign industries, as well as international scientific bodies, like the European Space Agency, have used the area of Quirra for their experiments.

From the very beginning, the Quirra test range has been regarded as a strategic site, due to its location and features. Today, it still has a great importance, because, according to the Air Force Staff, it is particularly suitable for the so-called military 'hot activities', namely training activities carried out by using exercise munition (both projectile and mortar fire). Apart from being used as a 'war games' area, both by Italian and foreign forces, the 'Salto di Quirra' has been used by military and arms manufacturers to test new bullets, rockets, and laser guided precision bombs as well as for the setting up of war planes, helicopters and drones. Some industries have also been experimenting with the effects of gas and oil pipelines explosions.

1 See L. Peruzzi, *Salto di Quirra, Il poligono sperimentale e di addestramento interforze compie 50 anni*, in *Rivista marittima*, Marina militare, 2006; *Convenzione tra il Ministero della Difesa ed il Dipartimento di Scienze Ambientali 'G. Sarfatti' dell'Università di Siena per lo studio geochimico-ambientale dell'area della Sardegna sud-orientale su cui insistono i Poligoni Militari di Perdasdefogu e Capo San Lorenzo, Relazione finale*, 15 June 2004, p. 8, <www.difesa.it/smd_/approfondimenti/poligonoperdasdefogu/documents/64919_relazione_finale.pdf>.

Even though military authorities affirm that the areas used for experiments are recurrently cleaned up, it is hardly surprising that several parts of the territory have seriously deteriorated. Moreover, one must consider that activities in the test range started at the end of the Fifties, when the sensibility for environmental concerns was presumably much more superficial than nowadays. This has been confirmed by a recent survey of the territory, which has revealed that the clean-up of the land, when not completely omitted, has been carried out in a very rough way. Especially in the past, the military personnel used to simply bury in several sites of the range the residual products of training activities and experiments, together with many kinds of waste and obsolete arms.

Since 2001, mass media have started talking about a real 'Quirra Syndrome',[2] in analogy with the 'Gulf War Syndrome' and the 'Balkan Syndrome', while referring to the abnormal percentage of haemolymphatic tumours (particularly leukaemia) reported among people living in the areas closest to the test range, as well as to cases of animals with severe birth defects.

5.2 The Judicial Investigation

In January 2011, the Italian judiciary (Prosecutor's Office of Lanusei) launched an investigation regarding the environmental situation of the areas around the Quirra test range, which ended on 25 March 2012 with the indictment of twenty people, including the former commanders of the range. On 12 March 2013, in order to decide on the indictment request formulated by the Prosecutor, the Judge of Lanusei ordered a new assessment of the condition of the area. On 18 December 2014 the trial was adjourned in order to wait for the ruling of the Constitutional Court on a preliminary question. For the purposes of the present contribution, it suffices to report the following findings of the Prosecutor's investigation:[3]

- a lot of military waste was found both in the land area and in the sea;
- between 1984 and 2008, explosions to destroy military waste were carried out without appropriate preventive measures to protect operators as well as the environment;
- toxic materials, like ballistite, were found in the area;

2 Senato della Repubblica, *Commissione parlamentare di inchiesta sui casi di morte e di gravi malattie che hanno colpito il personale italiano impiegato all'estero, nei poligoni di tiro e nei siti in cui vengono stoccati munizionamenti, in relazione all'esposizione a particolari fattori chimici, tossici e radiologici dal possibile effetto patogeno, con particolare attenzione agli effetti dell'utilizzo di proiettili all'uranio impoverito e della dispersione nell'ambiente di nanoparticelle di minerali pesanti prodotte dalle esplosioni di materiale bellico e a eventuali interazioni*, XVI Legislature, <www.senato.it/service/PDF/PDFServer/DF/288867.pdf>.

3 The principal findings of the judiciary investigation have been described by the Prosecutor of Lanusei, Domenico Fiordalisi, during the hearing of 8 May 2012 of the so-called *depleted uranium*'s Commission of inquiry (see Senato della Repubblica, *Commissione parlamentare di inchiesta sui casi di morte e di gravi malattie*, above n. 2, p. 64 et seq.)

- the Italian Regional Environmental Protection Agency also found higher levels of heavy metals in the zones of the explosions, thus revealing a risk of exposure to toxic substances;
- a landfill site of hazardous materials (asbestos, electronic apparatuses , tyres, batteries, igniters for missiles, components of anti-tank missiles) was discovered at 'Is Pibiris', near a tributary of the Flumendosa river;
- thorium-232 was found in the bodies of 12 shepherds who died of cancer. Thorium-232 is a radioactive metal, which emits particles that are six times more hazardous for human health than those released by depleted uranium, with the peak of toxicity between 20 and 25 years after use; thorium-232 was used as optical tracer of the anti-tank missile 'Milan', employed at the firing range between 1986 and 2000.

5.3 The Case of the 'Quirra' Test/Training Range under the EU/Euratom Law

5.3.1 *Parliamentary Questions to the EU Commission*

Besides the judiciary investigation, the situation surrounding the Quirra Test Range has been the subject of several questions put to the European Union Commission by some members of the EU Parliament, under Article 230.2 of TFEU. Regrettably, the Commission has not showed to have the will to go into details of the case under consideration, as well as of similar cases which have been the subject of other parliamentary questions.

For instance, on 20 January 2011 a question titled 'Cancer incidence in the areas surrounding the inter-force military test site in Quirra'[4] was put, referring to 'an abnormal percentage of haemolymphatic tumours' among people living in areas around the range, noticed by a research carried out by the health authorities of Cagliari and Lanusei. Moreover, as mentioned in the same parliamentary question, animals with severe birth defects had also been reported in those areas and tests carried out by the University of Modena had found 'traces of heavy metal nanoparticles' identical to those found in Italian soldiers who had died of cancer after serving in the Balkans. Particularly, considering that during the 1980s thirteen cases of birth defects had been reported in Escalaplano, a town of around two thousand people near to the range, the Commission was being requested to ascertain whether the Italian authorities, by continuously delaying the investigations on possible links between the use of experimental weapons in the test site and cancer incidence among

4 Uggias, *Question for written answer to the Commission (E-000181/2011)*, Parliamentary questions, 20 January 2011.

population, were infringing 'the citizens' fundamental right to health and environmental salubrity'.[5]

The answer given by the Commission was that 'according to the Treaty on the functioning of the European Union, the organisation of health systems and the delivery of healthcare is the responsibility of Member States. The Commission has no competence to require the Italian Authorities to perform an epidemiological study on clustering of negative health effects in the vicinity of a national military test site. The Commission does not intend to analyse the possible existence of a local conflict of interest or the violation of the fundamental right to public health and environmental health. Should the conditions in Quirra outside the military site move to breach European laws protecting human health, the Commission would ask the Italian Authorities for remedial action'.[6]

At first sight, one should admit that such an answer is formally correct: indeed, Article 168.1 of TFEU only provides that Union action on public health 'shall *complement* national policies'[7] and paragraph 7 of the same Article reads precisely that 'Union action shall respect the responsibilities of the Member States for the definition of their health policy and for the organisation and delivery of health services and medical care'. The Commission seems to refer to such a provision, in order to exclude its own competence to require the Member State's Authorities to perform an epidemiological study on health effects of military activities. But actually this is not what the question was asking for. Indeed, it was quite evident that an issue of *providing the public with relevant environmental information* was the core of the problem addressed in the latter, rather than of health systems and of delivery of healthcare. Particularly, the aforementioned parliamentary question had pointed out that the investigations on the environmental conditions around the military site were being coordinated by the Ministry of Defence, that is to say the same Authority under which the activities accused of environmental pollution had been carried out. Clearly, such a circumstance has not passed unnoticed to the Commission, which has expressly refused 'to analyse the possible existence of a local conflict of interest'. But such a refusal also means that the Commission has knowingly accepted the possibility that the truth, regarding environmental conditions of the areas and their potential negative health impact, never be disclosed to the public. In view of this, it even seems hypocrite the conclusive statement that the Commission would ask the Italian Authorities for remedial action only in case 'the conditions in Quirra outside the military site move to breach European laws protecting human health'. Indeed, as will be stressed in the following remarks, such an attitude by the Commission seems to disregard the fact that, among the actions by which the EU

---

5 Idem.

6 *Answer given by Mr Dalli on behalf of the Commission (E-000181/2011)*, Parliamentary questions, 3 March 2011.

7 Emphasis added.

protects human health, the Union policy on the environment has obviously a leading position.

On 23 April 2012, another question, concerning the 'Pollution caused by military activities in areas surrounding the Quirra firing range',[8] was put to the Commission by the same Parliament's member who had questioned it the year before. With reference to the worrying findings of the judiciary investigation concluded by the Public Prosecutor of Lanusei, as well as to the previous Commission's promise to ask the Italian Authorities for remedial action in case of environmental conditions amounting to a breach of EU law protecting human health, while complaining about the violation of 'the fundamental right of citizens to health, including that of the environment', this time the question addressed the issue of the clean-up of the areas polluted by the military activities and of possible penalties to be imposed in accordance with the European 'polluter pays' principle.

Once again, the Commission's answer has been formally correct, but, at the same time, its grounds for refusal to follow the case are not convincing:

> Directive 2004/35/EC on environmental liability with regard to the prevention and remedying of environmental damage (ELD – Environmental Liability Directive) aims at preventing and remedying environmental damage, based on the polluter-pays principle. (...) *Military activities are not covered* by occupational activities dangerous to the environment listed in Annex III to the ELD which fall under strict liability. Moreover, the ELD *excludes expressly activities 'the main purpose of which is to serve national defence or international security'.* Therefore, *the Environmental Liability Directive is not applicable to the pollution caused by military activities* in areas surrounding the Quirra firing range.[9]

Indeed, the aforementioned exemption grounded on 'national defence' or 'international security', as well as similar exemptions provided in other EU legislation concerning environmental issues, would be worth being assessed in a larger context, bearing in mind not only the single European legislative act under consideration, but also any relevant rule of the TEU and of TFEU, starting from the Title XX of the latter, concerning the Union policy on the environment. Anyway, because such a suggested larger assessment might also apply to other analogous arguments used by the Commission in answering to further parliamentary questions, always in order to justify its inaction with regard to the *Quirra* case or to similar cases, it seems better to mention other Commission's answers, before taking an overall view of them. Then, after such a comprehensive assessment, it will be even possible

8 Uggias, *Question for written answer to the Commission (E-004169/2012)*, Parliamentary questions, 23 April 2012.

9 *Answer given by Mr Potočnik on behalf of the Commission (E-004169/2012)*, Parliamentary questions, 19 July 2012, emphasis added.

to attempt a reply to another formally correct Commission's statement, particularly that 'the EU does not have general competency in the area of fundamental rights. The European Commission is not empowered to assess whether the fundamental rights have been respected by a Member State when the alleged breach of fundamental rights is not linked with the application of EC law'.[10]

Environmental concerns linked to military activities carried out in Sardinia, particularly around the Quirra test range, had been also the subject of another parliamentary question, titled 'Military bases in Sardinia', on 3 April 2012. Here, the findings relating to the numerous cases of cancer and leukaemia, among the residents of the areas concerned, had been expressly reported as a result of a situation of high pollution with 'radioactive materials, industrial heavy metal particles, depleted uranium, thorium-232 and cadmium'[11] within the military areas used for NATO activities. Hence, the Commission had been requested whether it was aware of the negative and dangerous effects on the soil, air and water and what it had already done to investigate the accuracy of the information provided, to monitor the condition of the area and to urge Italian authorities to decontaminate it.[12] In addition, in the parliamentary question under consideration reference had been made to other 'similar hidden natural disasters in Europe, especially in Austria, which have been caused by military experiments and weapons testing and could affect the health of EU citizens'.[13]

In view of the reference to the radioactive pollution, this time it has been even too easy, for the Commission, in order to justify its refusal to investigate the case and to urge Italian authorities, to invoke the Court of Justice's ruling that 'the Euratom Treaty and its secondary legislation do not apply to activities or practices of military nature'.[14] It has also answered that

> the monitoring of the level of radioactivity in the air, water and soil (…) is a national responsibility, while it is also the responsibility of the Member State concerned to ensure compliance with the applicable safety standards (including dose limits for members of the public). The Commission has no evidence,

10 *Answer given by Mr Potočnik on behalf of the Commission (E-004169/2012)*, Parliamentary questions, 19 July 2012.

11 Werthmann, *Question for written answer to the Commission (E-003560/12)*, Parliamentary questions, 3 April 2012.

12 Idem, points 1-3.

13 Idem, point 4.

14 Reference is made to the Judgment of 9 March 2006, in Case C-65/04, *Commission v. UK*. This Judgment, indeed, has essentially confirmed the previous Judgment of 12 April 2005, in Case C-61/03, *Commission v. UK*. For a critical comment on such Judgments, see Chapter 2 in this volume by Sousa Ferro.

> according to the information it receives from Member States that the area referred to by the Honourable Member is contaminated.[15]

The same reasoning has been applied by the Commission to further parliamentary questions alleging similar cases of suspected radioactive pollution linked to military activities. For example, on 20 December 2012, a question titled 'Use of radioactive weapons in military activities and risk to human health in the EU', addressed the issue of the death of several Italian soldiers by tumoral diseases, apparently related to radioactive contamination resulting from the use of depleted uranium weapons, while serving in military missions in the former Yugoslavia or operating within Italian firing ranges where arms tests are carried out. In such a context, among other things, the Commission was also requested whether it agreed or not that 'it might be appropriate to monitor and study the impact that activities carried out on EU military firing ranges might have on human and animal health, groundwater and the surrounding environment'.[16]

Apart from invoking, once again, the aforementioned Court's ruling on the Euratom Treaty and pedantically repeating that the monitoring of the level of radioactivity in the air, water and soil is a national responsibility, the Commission has expressed its 'view' on the issue of the possible impact on human health and environment from the use of depleted uranium weapons:

> As far as depleted uranium ammunition is concerned, there is an ongoing debate on its alleged impact on human health and environment. On the basis of studies conducted by specialised international agencies so far, there is still no scientifically demonstrated link between the use of depleted uranium in operational theatres and damage to human health. However, some of these reports and other authoritative studies have underlined the need to continue to monitor the long-term effects of depleted uranium on human beings and the environment.[17]

---

15 *Answer given by Mr Oettinger on behalf of the Commission (E-003560/12)*, Parliamentary questions, 25 May 2012. Moreover, by such an answer it has been specified that '[t]he Commission's REM (Radiological Environmental Monitoring) database does not contain any values for uranium (or another radionuclide relevant to this matter) in environmental samples from Sardinia. Within the Commission's verification schedule under Article 35 of the Euratom Treaty and in continuation of work started in 2010, a visit to Sardinia (and other Italian regions) is foreseen for 2013 with a view to verifying the state of the monitoring facilities for environmental radioactivity in these areas.'

16 Zanoni, *Question for written answer to the Commission (E-011684-12)*, Parliamentary questions, 20 December 2012.

17 *Answer given by Mr Borg on behalf of the Commission (E-011684-12)*, Parliamentary questions, 25 February 2013.

On the one hand, such an answer seems to pass over the suggestion to monitor the impact of activities carried out within EU military firing ranges (and not only those activities implying the use of ammunition releasing radioactive substances) on human and animal health, groundwater and the 'surrounding environment', which last expression had been evidently used to exclude from the call any Commission's monitoring role *inside* military firing ranges. On the other hand, the conclusive remark on the depleted uranium issue seems to describe the typical situation to which the precautionary principle applies,[18] though without the Commission deducing the consequences one could expect from it.

More recently, the Commission has been offered a new occasion to confirm more expressly its 'renunciatory' attitude towards the issue of environmental impact of activities carried out inside military areas in Europe, also outlining the strict conditions that it considers as insurmountable limits to its possible intervention in the matter. On 6 January 2014, a question titled 'Alarming levels of thorium-232 at the military firing range lying between Cordenons, San Quirino, Vivaro and San Giorgio della Richinvelda, in the province of Pordenone' was put to the Commission.[19] In late December 2013, the results of tests carried out by the Friuli-Venezia Giulia provincial department of the Italian Regional Environmental Protection Agency had showed alarming levels of thorium-232 in the area of the aforementioned firing range: indeed, between 1986 and 2003 the famous 'Milan' anti-tank missiles (emitting thorium-232) had been used within the area, during military training activities. As recalled in the text of the parliamentary question, the same missiles had been also used at the Quirra firing range. Moreover, near the Pordenone's firing range, also the 'Magredi' region, an area protected as both a site of Community importance and a Special Protection Area within the meaning of the Habitats Directive[20] and the Birds Directive,[21] was at risk of contamination. In sum, the Commission was asked what initiatives it intended to implement in order to prevent similar cases of pollution in the EU, and moreover to prevent the contamination of aquifers.

As usual, the Commission has replied that it has 'received no information on specific cases since military activities are not covered by the reporting obligations under Articles 35 and 36 of the Euratom Treaty. (...) [A]s the Commission's competence under the Euratom Treaty does not extend to military activities, it cannot take any initiatives on

18 Suffice it to refer to the ECJ description of the principle: 'Where there is uncertainty as to the existence or extent of risks to human health, the institutions may take protective measures without having to wait until the reality and seriousness of those risks become fully apparent' (Case C-180/96, *UK v. Commission*, Judgment of 5 May 1998, §99).

19 Zanoni, *Question for written answer to the Commission (E-000031-14)*, Parliamentary questions, 6 January 2014.

20 Council Directive 92/43/EEC of 21 May 1992 on the protection of natural habitats and wild fauna and flora, OJ L 206, 22.07.1992.

21 Directive 2009/147/EC.

military sites'.[22] But what it is worth stressing here is the following view that '*Outside* military areas, the monitoring requirements laid down in Article 35 of the Euratom Treaty and the Basic Safety Standards Directive apply'.[23] In addition to this, the Commission has also assured that it monitors closely the implementation of the Water Framework Directive,[24] as well as of the Directive 2013/51/Euratom.[25]

5.3.2 *The ECJ's Opinion that the Euratom Treaty Does Not Apply to Military Activities as a Justification of the Commission's Failure to Intervene*

It is self-evident that the aforementioned Court's Judgments of 2005 and 2006, affirming that the Euratom Treaty does not apply to military activities, has left a 'stumbling block' on the Commission's shoulder: as a result, whenever it hears the word 'military', it seems automatically to surrender and believe that it cannot do anything, even in cases where the risk of environment pollution is well-founded. The other negative consequences of the same Judgments and the latter's inconsistency with the previous Court's case law are well discussed in another part of the present book[26] and do not need to be examined in depth here too. Suffice it to make the following remarks.

What seems far from convincing, within the Court's reasoning, is not its conclusion that the Euratom Treaty 'is not applicable to uses of nuclear energy for military purposes': indeed, in its literal sense, one could even agree with such an affirmation, insofar as uses of nuclear energy in the military sector, since the beginning of their development, have always been a very sensitive States' interest, which they usually manage in all secrecy. Just to give an example, when a State invests its resources in research to produce a nuclear weapon, or a nuclear system of propulsion for its warships, after all it aims at getting an advantage over other possible and future enemy States. How could one imagine States binding themselves, by treaty, to share the results of such a research in the military sphere, thus forgoing their advantage over the others? It seems realistic raising doubts, at least if one considers the particular historical context at the time of conclusion of the Treaty.

However, what rather seems unforgivable in the Court's reasoning is the logic leap which ultimately leads it to the further conclusion that even *the possible effects* of uses of nuclear energy for military purposes, namely their negative impact on human health and

22 *Answer given by Mr Oettinger on behalf of the Commission (E-000031-14)*, Parliamentary questions, 26 February 2014, §1 and §3.
23 Idem, §3, emphasis added.
24 Directive 2000/60/EC of the European Parliament and of the Council of 23 October 2000 establishing a framework for Community action in the field of water policy, OJ L 327, 22.12.2000.
25 Council Directive 2013/51/Euratom of 22 October 2013 laying down requirements for the protection of the health of the general public with regard to radioactive substances in water intended for human consumption, OJ L 296, 07.11.2013.
26 See Sousa Ferro, above n. 14.

environment, are by definition *outside* the scope of the Euratom Treaty. Such a conclusion is more evident in the Court's view in the case relating to the failure by the United Kingdom to provide the Gibraltar public with information on protection measures in the event of a radiological emergency.[27] In such a case, only because the possible source of radiological emergency was a Royal Navy nuclear submarine berthed in Gibraltar harbour and in order to confirm its previous Judgment of 2005,[28] the Court has eventually stated that Euratom secondary legislation cannot be applied to *any* situation connected with the military sphere. Indeed, the applicability of Article 5(3) of Council Directive 89/618/Euratom[29] was at issue. Whereas, in the Court's view,

> the use of nuclear energy for military purposes falls outside the scope of all the provisions of the EAEC Treaty, not just some of them. (…) As the scope of provisions of secondary legislation cannot validly exceed that of their legal basis, the inapplicability of Article 31 EA to military activities necessarily means that the directive does not apply to such activities.[30]

Such reasoning seems to disregard the general rule of interpretation laid down in Article 31 of the Vienna Convention on the Law of Treaties. As the Court itself has held in its case law, several provisions of such a Convention reflect customary international law of treaties, which, as such, binds both the Community and Member States, independently of the same Convention.[31] This is certainly the case of the aforementioned Article 31 of Vienna Convention, under which no treaty, however special its subject-matter, can be interpreted and applied as if it existed in a legal vacuum. Particularly, it must be recalled that a treaty shall be interpreted also in its context, which comprises 'any agreement relating to the treaty which was made between all the parties in connection with the conclusion of the treaty'[32] and, moreover, taking into account 'any relevant rules of international law applicable in the relations between the parties'.[33] From this point of view, one could argue that it is not a matter of respecting the Euratom Treaty's nature of *lex specialis*,[34] together with its restricted scope according the Court's interpretation. One must rather realize that there are some rules, both customary and substantially embodied in the Euratom connected

27 Judgment of 9 March 2006, in Case C-65/04, *Commission v. UK*.

28 Judgment of 12 April 2005, in Case C-61/03, *Commission v. UK*.

29 Council Directive 89/618/Euratom of 27 November 1989, on informing the general public about health protection measures to be applied and steps to be taken in the event of a radiological emergency, OJ 1989 L 357.

30 Judgment of 9 March 2006, in Case C-65/04, *Commission v. UK*, §§26 and 27.

31 See, for instance, Judgment of 25 February 2010, Case C-386/08, *Brita*, §§40-43.

32 Vienna Convention on the Law of Treaties, Art. 31, §2(a).

33 Vienna Convention on the Law of Treaties, Art. 31, §3(c).

34 Judgment of the Court of First Instance 6 May 2009, *Outokumpu Oyj et al. v. Commission*, Case T-122/04, §55.

treaties (TEU/TFEU), which have always to inspire the conduct of all the Community Institutions and of Member States. Even under the *special* Euratom legal system, those same rules may not tolerate derogations which ultimately might lead to a prejudice of the *collective* interests they are aimed at protecting. Particularly, reference is made here to the rules protecting both human health and the environment. Indeed, these latter are typical examples of 'indivisible interests': as well-known, under international law, the rules established for their protection are regarded as a special category, namely *erga omnes* obligations, which, according to the definition adopted here,[35] are intended as customary law or multilateral treaty obligations that, in view of the indivisibility of their particular content, bind simultaneously all States (in case of rules of general international law) or all the agreement's states parties (in case of multilateral treaty rules). It is not argued here that any derogation to similar rules must *per se* be regarded as *void*. Such a legal effect might follow only if the peremptory nature of the same rules was demonstrated. Rather it is simply held that, even if one grants, for the sake of argument, that that Euratom Treaty and its secondary legislation (as well as EU law) could be interpreted as implicitly derogating, *inter omnes partes*, the aforementioned *erga omnes* obligations (particularly whenever Member States act within the military sphere or for defence purposes), in case of actual infringement of the customary rules protecting the environment and human health, a Member State should be regarded as responsible towards third States. This effect follows precisely from the *erga omnes* nature of the rules at issue[36] and ultimately explains why any treaty obligation should always be interpreted and applied in accordance with such a special category of norms.

In view of all the considerations above made, it seems that also the Euratom Treaty should be interpreted taking into account 'any relevant rules of international law applicable in the relations between the parties'. Among such a relevant rules, under customary law,

35 G. Arangio-Ruiz, *Fourth report on State responsibility*, Extract from the Yearbook of the International Law Commission, 1992, Vol. II(1), A/CN.4/444 and Add. 1-3, §92: 'the concept of *erga omnes* obligations is not characterized by the importance of the interest protected by the norms – this aspect being typical of *jus cogens* – but rather by the 'legal indivisibility' of the content of the obligation, namely by the fact that the rule in question provides for obligations which bind simultaneously each and every addressee with respect to all others. This legal structure is typical not only of peremptory norms, but also of other norms of general international law and of a number of multilateral treaty rules (*erga omnes partes* obligations)'.

36 In a similar vein, see *Report of the Study Group of the International Law Commission (finalized by Martti Koskenniemi), Fragmentation of International Law: Difficulties Arising from the Diversification and Expansion of International Law*, International Law Commission, Fifty-eighth session (Geneva, 1 May-9 June and 3 July-11 August 2006), A/CN.4/L.682, 13 April 2006, p. 83, §154: 'there is little doubt that most international law (…) is dispositive and that contracting out by establishing a regime is possible and limited only to the extent that such limitation may be received from the *jus cogens* nature or otherwise compelling character of general law. Aside from peremptory norms, at least the following limitations should be considered: (…) (2) *The regime may not deviate from general law if the obligations of general law* are of 'integral' or 'interdependent' nature, *have erga omnes character* or practice has created a legitimate expectation of non-derogation' (emphasis added).

one could not disregard the States' duty to act with *due diligence* to prevent, reduce and control not only transboundary harms to the environment of other States or to the so-called 'global common areas', but even to their own *domestic* environment, at least those environmental harms so serious as to actually or even potentially affect human health. In the light of such a duty *erga omnes* to protect the environment, which ultimately also aims at protecting human health,[37] it is arguable that the Euratom's objective to protect the health of workers and of the general public against the dangers arising from ionizing radiations acquires nowadays a more pregnant content. Indeed, Euratom's activities to pursue that objective may be regarded as a way by which Members States aim at complying, both collectively and individually, with their duty to protect the environment. In other words, the protection against the dangers arising from ionizing radiations must be considered as a Member States' environmental policy's part, which they mostly manage by the means established under the Euratom Treaty.

Just to give an example of how the latter has already been interpreted in the light of the duty to protect the environment, eventually extending the original scope of the action against the dangers arising from ionizing radiations, one may refer to Council Directive 2013/59/Euratom of 5 December 2013, point 27 of the preamble:

> The contamination of the environment may pose a threat to human health. The Community's secondary legislation *so far* has regarded such contamination *only as a pathway of exposure to members of the public directly affected by radioactive effluent discharged to the environment. While the state of the environment can impact long-term human health, this calls for a policy protecting the environment against the harmful effects of ionising radiation.*[38]

Such a statement seems emblematic of the new environmental approach by which Member States are interpreting the Treaty at issue, while adopting the related secondary legislation, though that Treaty has substantially remained unchanged since 1957 and still does not even contain the word 'environment'.

Moreover, among the relevant rules of international law applicable in the relations between the Member States, one must take into account also the TEU and the TFEU, as all Member States are also Parties to the Euratom Treaty. In this context, the Euratom's action against the dangers arising from ionizing radiations may be also regarded as a part of the EU policy on the environment, to which shared competence between the Union

37 Judgment of 21 December 2011, Case C-28/09, *Commission v. Austria* §122: 'The objective of protection of health is therefore already incorporated, in principle, in the objective of protection of the environment.'

38 Council Directive 2013/59/Euratom of 5 December 2013, laying down basic safety standards for protection against the dangers arising from exposure to ionizing radiation, and repealing Directives 89/618/Euratom, 90/641/Euratom, 96/29/Euratom, 97/43/Euratom and 2003/122/Euratom, emphasis added.

and the Member States applies (Article 4.2(e) of TFEU). This is not without important consequences from the interpretative point of view, because it means that the action against radioactive pollution should also pursue the objectives of the Union environmental policy, such as, among others, preserving, protecting and improving the quality of the environment and protecting human health (Article 191.1 of TFEU). Furthermore, under Article 191.2 of TFEU, Union policy on the environment 'shall aim at a high level of protection' and 'shall be based on the precautionary principle and on the principles that preventive action should be taken, that environmental damage should as a priority be rectified at source and that the polluter should pay'. It seems worth referring also to Article 3.3 of TEU, under which '[t]he Union (...) shall work for the sustainable development of Europe based on (...) a high level of protection and improvement of the quality of the environment', as well as to Article 11 of TFEU: 'Environmental protection requirements must be integrated into the definition and implementation of the Union's policies and activities, in particular with a view to promoting sustainable development.'

To turn back to the Euratom Treaty, for the sake of clarity, here it is not suggested that the same agreement should be applied, wholly or partially, to 'military activities', or even to 'the use of nuclear energy for military purposes'. What is rather held here is that, taking into account all the relevant rules of international law applicable in the relations between the parties, the Euratom provisions aimed at protecting the environment against the harmful effects of ionising radiation, as well as the related secondary legislation, should be interpreted and applied within a larger legal context, which is also provided by Title XX of the TFEU, and so under the 'shared competence' scheme of the Union policy on the environment. Within such a larger context, all applicable means offered by the Euratom Treaty to fight against radioactive pollution should come into play, *whatever be the source (civilian or military) of such a pollution.* Of course, this does not mean that *any* Euratom rule relating to the protection of the environment from the harmful effects of ionising radiation might always be applicable *as such*, even when the possible source of contamination is of a military nature. Particularly, it is plain that in similar cases certain consultative or 'police' Commission's powers, provided by the Treaty, should be applied only to the extent that Member States' essential security interests are not jeopardised. Obviously, it would be a matter of striking a balance between, on the one hand, environmental and health concerns and, on the other hand, State's national security interests, on a case-by-case basis. Indeed, such a solution has been suggested even by the Advocate General Geelhoed, in his opinion in Case C-61/03, relating to the decommissioning operations of a UK military nuclear reactor.[39] What's more, it is interesting how the Advocate General has interpreted Euratom Treaty taking into account the treatment of Member State's

39 Case C-61/03, *Commission v. UK, Opinion of Advocate General Geelhoed*, 2 December 2004, §108 et seq.

defence interests under the EC Treaty and the relationship between these agreements.[40] Particularly, the Advocate General's opinion seems to suggest that, even though the Euratom Treaty does not contain a general clause aimed at preserving Member State's essential security interests as regards the supply of defence-related information, like Article 296 of TEC (corresponding to the new Article 346 of TFEU),[41] nonetheless the same latter provision might well be invoked by Member States also to protect their essential defence interests *under the Euratom Treaty.*[42] In substance, the Advocate General has suggested to the Court the same interpretation as suggested here, which calls for the taking into account all the relevant rules of international law applicable in the relations between the parties, as provided by Article 31, § 3(c) of the Vienna Convention.

That being stated, as above recalled, also the Euratom's action to protect the environment against the harmful effects of ionising radiation must be considered within the purview of the Union's policy on the environment and as an area in which the Treaties confer on the Union a competence shared with the Member States. Particularly, it is arguable that, with regard to the *specific area* of the protection against radioactive pollution arising from *military* sources, the competence has remained with the Members States. But this does not mean that, *within such a specific area*, Members States are totally free to act in contrast with the action, carried out by the Union and the Euratom, against the harmful effects of ionising radiation from *civilian* sources. Indeed, they are bound under their general duty of loyal or 'sincere' cooperation, provided both by Article 192 of Euratom Treaty and by Article 4.3 of TEU, to facilitate the achievement of the Euratom/Union's tasks and refrain from any measure which could jeopardise the attainment of the Euratom/Union's objectives. From this point of view, it seems plain that Member States, in the specific area of radiological risks from military sources, could not exercise their competence on the environment's protection in such a way as to jeopardize the general Euratom/Union objective to protect

40 Case C-61/03, *Commission v. UK, Opinion of Advocate General Geelhoed*, §§100-102 and §§104-105.

41 Art. 346 of TFEU (*ex* Art. 296 TEC) reads: '1. The provisions of the Treaties shall not preclude the application of the following rules: (a) no Member State shall be obliged to supply information the disclosure of which it considers contrary to the essential interests of its security; (b) any Member State may take such measures as it considers necessary for the protection of the essential interests of its security which are connected with the production of or trade in arms, munitions and war material; such measures shall not adversely affect the conditions of competition in the internal market regarding products which are not intended for specifically military purposes. (...)'

42 Case C-61/03, *Commission v. UK, Opinion of Advocate General Geelhoed*, §107 (emphasis add): '(...) the relationship between the EA and EC Treaties belies the existence of any "lacuna" stemming from the absence in the EAEC Treaty of an analogous provision to Article 296 EC. Rather, insofar as the EAEC Treaty and implementing legislation do not explicitly provide for equivalent safeguards for Member State's essential security interests, the safeguards contained in Article 296 EC may equally apply to products covered by the EAEC Treaty.' Unfortunately, as well-known, the Court has not agree with the Advocate General's opinion. On the contrary it has accepted the UK argument that the very absence in the Treaty of provisions like the aforementioned Art. 296 of TEC 'leads to the conclusion that activities falling within the military sphere are outside the scope of that Treaty' (Judgment of 12 April 2005, in Case C-61/03, *Commission v. UK*, §36).

the environment against the harmful effects of ionising radiation, as well as the objectives to improve the quality of the environment and protect human health, which moreover constitute the indivisible content of customary *erga omnes* international obligations. The same reasoning, of course, applies to the objectives pursued by Euratom's secondary legislation on the protection of human health and environment.

After all, on the one hand, one could blame the Commission for having attempted, even too strictly, to apply *as such* Article 37 of Euratom Treaty in the case of the UK military nuclear reactor's decommissioning operations, failing to find a solution with the UK Government, which could have allowed the latter to supply information on the disposing activities of military radioactive waste, while at the same time withholding certain information to preserve its essential security interests.

On the other hand, even today, the Commission's refusal to investigate how Members States manage the issue of the protection of the environment from possible radioactive contamination caused by military activities seems to be a disproportionate reaction to the aforementioned Court's Judgments. Moreover, such a 'renunciatory' attitude seems to bear out the unacceptable idea that the protection of the environment might be regarded as a divisible interest, which should in principle be superseded by the Member States' defence interests. On the contrary, it seems arguable that, in principle, the latter State's interests should be superseded by the indivisible objective of the environment's protection. Indeed, suffice it to recall that, as above stressed, environmental protection requirements are aimed at promoting *sustainable development*. That said, one must emphasize that the State military activities at issue here are to be regarded as *functional* to a legal use of military force. In other words, they are *preparatory activities for war* in a self-defence context. According to Principle 24 of the Rio Declaration on Environment and Development, 'Warfare is inherently destructive of sustainable development'. Hence, when striking the balance between environmental concerns and State defence interests, one should take into account that also those preparatory military activities could even have a 'destructive' impact on the objective of sustainable development.

5.3.3 *The Protection of the Environment against Harmful Effects of Military Activities under TEU/TFEU and Secondary Legislation*

The aforementioned Commission's attitude seems even more unjustified when one consider the case at issue, as well as other similar cases, under EU primary and secondary law. Just as a starting point, suffice it to refer again to other Commission's answers to some parliamentary questions.

On 3 September 2012, a question titled 'New US military base in the province of Vicenza: lack of transparency in regard to citizens and possible environmental implications'[43] was put to the Commission. The subject of the claim was the Italian Government's decision to authorize the construction of a new US military base in the territory of Vicenza, near a site of Community importance without either informing the public or involving it in the decision-making. The Commission was requested to check whether the decision had complied with all procedural requirements, including the Aarhus Convention of 25 June 1998 on Access to Information, Public Participation in Decision-Making and Access to Justice in Environmental Matters, to which also the EU is itself a party.

According to the answer given by the Commission, '[a]s for the Aarhus Convention, (…) the establishment of a military training centre is to be regarded as *an activity serving national defence purposes*. As such, and in conjunction with national law, this activity is outside the scope of the Aarhus Convention'.[44]

The same Aarhus Convention was also invoked more recently, on 31 October 2014, by another parliamentary question, titled 'NATO military bases in Sardinia'.[45] After making reference, once again, to the well-known complaints of extensive damage to environment and contamination caused by the activities carried out in the military bases in Sardinia, the Commission was requested to assess whether Italy has complied with the Aarhus Convention requirements, particularly *whether it has provided its citizens with the relevant environmental information*. This last part of the question concerns a very important aspect of the *Quirra* case and of other similar cases, as will be discussed below. Thus, it seems even surprising the following laconic answer given by the Commission rather: 'The Commission is not aware of problems related to the implementation of the Aarhus Convention in Italy as concerns NATO military bases in Sardinia.'[46]

Moreover, still relating to a matter of environmental information, it seems disconcerting the answer given by the Commission to a previous parliamentary question, which is worth

43 Bizzotto, *Question for written answer to the Commission (E-007788/2012)*, Parliamentary questions, 3 September 2012.

44 *Answer given by Mr Potočnik on behalf of the Commission (E-007788/2012)*, Parliamentary questions, 5 November 2012, emphasis added. In addition, the Commission answered that 'The same principle applies with regard to the application of Directive 2011/92/EU (Environmental Impact Assessment or "EIA Directive"). The EIA Directive grants discretion to Member States not to apply its provisions for *projects serving national defence purposes* (…). As regards the vicinity of the military training center to a site of Community importance (SCI), Directive 92/43/EEC ('Habitats Directive') *does not forbid military installations* or other projects in or near SCIs. It is up to the competent national authorities to assess whether such a project could cause significant negative effects on the relevant species and habitats of the concerned site and, if this is the case, to follow the rules and procedures set by Article 6 of the Habitats Directive' (idem, emphasis added).

45 Moi, *Question for written answer to the Commission (E-008628-14)*, Parliamentary questions, 31 October 2014.

46 *Answer given by Mr Vella on behalf of the Commission (E-008628/2014)*, Parliamentary questions, 5 January 2015.

mentioning here, even though it does not concern environmental issues arising from military activities. It was put to the Commission on 4 November 2013, under the title 'Concealment of massive amounts of hazardous and harmful toxic waste in Italy and non-disclosure of important environmental information by the Italian authorities'.[47] On 31 October 2013, the Bureau of the Italian Chamber of Deputies had published the minutes of a hearing, held on 7 October 1997, by the Parliamentary Committee of inquiry into the waste cycle and related illegal activities. During such a hearing, a mafia turncoat, Carmine Schiavone, had revealed the 'Camorra' clans' illegal and systematic practice of concealment of hazardous, even radioactive, waste, by burying it in the Italian regions of Campania, Lazio and Molise. Even the sites where waste was buried had been pinpointed by documents attached to the minutes at issue. As the question concludes, '[t]hese minutes contain extremely important environmental information which, if used immediately, could have limited health risks and reduced ground and water table pollution. Can [the Commission] state whether the Italian authorities' failure to disclose them for the past 16 years complies with EU rules on waste, transparency of environmental information, the protection of water and of health?'[48]

On this particular point, the Commission answered that '[t]he Commission is not able to establish whether late publication of the minutes from the hearing of the Parliamentary Committee has affected the efforts of the Italian Government to identify and clean the sites. *The delay in the publication does not constitute a breach of EU rules on waste or access to environmental information.*'[49]

As regards the issue of environmental information, it must be recalled that, even before the former EC became party to the Aarhus Convention,[50] Directive 2003/4/EC on public access to environmental information[51] was adopted on 28 January 2003. By Article 7.4, such a Directive provides that

> [w]ithout prejudice to any specific obligation laid down by Community legislation, Member States shall take the necessary measures to ensure that, *in the event of an imminent threat to human health or the environment*, whether caused by human activities or due to natural causes, *all information held by* or

47 Zanoni, *Question for written answer to the Commission (E-012426-13)*, Parliamentary questions, 4 November 2013.

48 Idem.

49 *Answer given by Mr Potočnik on behalf of the Commission (E-012426/2013)*, Parliamentary questions, 7 January 2014, emphasis added.

50 The EC is party to the Aarhus Convention since May 2005. *See* the Regulation (EC) No. 1367/2006 *on the application of the provisions of the Aarhus Convention on Access to Information, Public Participation in Decision-making and Access to Justice in Environmental Matters to Community institutions and bodies.*

51 Directive 2003/4/EC of 28 January 2003 on public access to environmental information and repealing Council Directive 90/313/EEC.

for *public authorities* which could *enable the public likely to be affected to take measures to prevent or mitigate harm arising from the threat is disseminated, immediately and without delay.*

Even though the applicability of such a provision to the case described in the last EU parliamentary question might be controversial (the Italian Parliamentary Committee's hearing at issue was held on 7 October 1997, while the undeniable threat to human health or the environment, caused by the illegal burying of hazardous waste, was still actual and persistent even in 2003), the Commission's view that the delay in the publication of the minutes does not constitute a breach of EU rules on access to environmental information seems far from convincing. Moreover, one should note that since 13 Jun 2001 Italy ratified the Aarhus Convention, which contains a provision very similar to the aforementioned Article 7.4 of the Directive 2003/4/EC.[52] It also seems worth observing that, insofar as the mafia turncoat had referred, among other things, to the illegal burying of *radioactive* waste, such an information, if timely disclosed, could have allowed to detect a possible situation of *radiological emergency*, even for the purposes of the above mentioned Directive 89/618/Euratom,[53] whose Article 6.1 reads 'Member States shall ensure that, when a radiological emergency occurs, the population actually affected is informed without delay of the facts of the emergency, of the steps to be taken and, as appropriate to the case in point, of the health-protection measures applicable to it.'

Anyway, apart from specific provisions of EU/Euratom secondary legislation, as well as of the Aarhus Convention, it is arguable that a States' duty to inform the public, without delay, on possible risks arising from exposure to a contaminated environment (whatever be the source of contamination) should be regarded as included in the already mentioned States' customary duty to act with *due diligence* to prevent, reduce or control damage and pollution to their own domestic environment susceptible of having negative effects on human health. Moreover, such a State's duty to provide the public with relevant environmental information, in case of potential risks for human health, is even regarded as enforceable, both within national legal systems and under certain treaties protecting human rights, by means of justiciable individual fundamental rights, such as, in particular, the

52 Aarhus Convention of 25 June 1998 on Access to Information, Public Participation in Decision-Making and Access to Justice in Environmental Matters, Art. 5.1(c): '1. Each Party shall ensure that: (...) (c) In the event of any imminent threat to human health or the environment, whether caused by human activities or due to natural causes, all information which could enable the public to take measures to prevent or mitigate harm arising from the threat and is held by a public authority is disseminated immediately and without delay to members of the public who may be affected.'

53 Council Directive 89/618/Euratom of 27 November 1989, above n. 29; such a Directive has been repealed by Directive 2013/59/Euratom of 5 December 2013, laying down basic safety standards for protection against the dangers arising from exposure to ionising radiation, which contains a provision on information to the members of the public likely to be affected in the event of an emergency (Art. 70.1) similar to the Art. 6.1 above reported.

right to life or the right to respect for private and family life. This point will be discussed below, under the next part of this contribution.

However, such a reference to the possible infringement of human rights, linked with cases of environmental contamination, now calls for an assessment of the above reported Commission's statement that '[t]he European Commission is not empowered to assess whether the fundamental rights have been respected by a Member State when the alleged breach of fundamental rights is not linked with the application of EC law'.

Indeed, such a statement seems to reflect a Court's settled case law, according to which 'the fundamental rights guaranteed in the legal order of the European Union are applicable in all *situations governed by European Union law, but not outside such situations*', as well as the observation that the Court itself 'has no power to examine the compatibility with the [EU's] Charter [of Fundamental Rights] of national legislation lying *outside the scope of European Union law*'.[54] Moreover, Article 51.1 of the same Charter ('The provisions of this Charter are addressed to (...) the Member States only when they are implementing Union law') is to be interpreted to mean that 'the requirement to respect fundamental rights defined in the context of the Union is only binding on the Member States *when they act in the scope of Union law*'.[55]

Nonetheless, it must also be recalled the further Court's statement that

> in a situation where action of the Member States is *not entirely determined by European Union law*, (...) national authorities and courts remain free to apply national standards of protection of fundamental rights, *provided that the level of protection provided for by the Charter*, as interpreted by the Court, *and the primacy, unity and effectiveness of European Union law are not thereby compromised.*[56]

That being stated, should one conclude that a case of actual, or even potential, environmental contamination is to be regarded as 'outside the scope of European Union law' on the sole ground that such a contamination results from military activities?

Once again, it must be stressed that the protection of the environment is a subject to which shared competence between the Union and the Member States applies. This means that, in that specific area, the Union and the Member States may legislate and adopt legally binding acts. Moreover, in that same area, the Member States may exercise their competence 'to the extent that the Union has not exercised its competence' (Article 2.2 of TFEU). Particularly, it may well occur that, within such a context of shared competence, the Union

54 Judgment of 26 February2013, in Case C-617/10, *Åkerberg Fransson*, §19, emphasis added.

55 Idem, §20, emphasis added.

56 Idem, §29, emphasis added.

abstains from legislating on special issues, for instance in order to respect certain Member States' interests, such as national security and defence. Indeed, even under Article 4.2 of TEU, 'national security remains the sole responsibility of each Member State' and in the same vein one might interpret those 'exemptions' provided in certain EU secondary legislation, often invoked by the Commission in its aforementioned answers to some parliamentary questions. Suffice it to recall the Environmental Liability Directive (2004/35/CE), whose Article 4.6 reads '[t]his Directive shall not apply to activities the main purpose of which is *to serve national defence or international security*'.[57] One could also refer to Article 1.3 of the Environmental Impact Assessment Directive (2011/92/EU): 'Member States may decide, on a case-by-case basis if so provided under national law, not to apply this Directive to *projects serving national defence purposes*, if they deem that such application would have an adverse effect on those purposes.'[58]

From analogy with what already explained relating to the Euratom's scope of application, similar exemptions may be interpreted to mean that, with regard to *specific aspects* related to the objective of environment's protection, the competence has remained with Member States. But this does not mean that the same exemptions are intended to authorize Member States to act totally in contrast with the overarching objective of the protection of the environment, neither with the specific objectives pursued by each relevant act of secondary legislation. Indeed, in the light of the shared competence applied to the area at issue and of the aforementioned 'exemptions', a situation of environmental pollution arising from military activities should be regarded as 'a situation where action of the Member States is not entirely determined by European Union law', namely a situation in which national authorities remain free to apply national standards of protection of fundamental rights, but without compromising the level of protection provided for by the EU's Charter of Fundamental Rights, as well as 'the primacy, unity and effectiveness of European Union law'.

Moreover, in the same vein, one could refer to the Court's opinion that

> although it is for Member States to take the appropriate measures to ensure their internal and external security, *it does not follow that such measures are*

57 Emphasis added.

58 Emphasis added. In the same vein, one might mention the Directive 2003/35/EC of 26 May 2003, providing for public participation in respect of the drawing up of certain plans and programmes relating to the environment and amending with regard to public participation and access to justice Council Directives 85/337/EEC and 96/61/EC, particularly Art. 2.4: 'This Article shall not apply to plans and programs designed for the sole purpose of serving national defence (...)'; the Directive 2011/65/EU of 8 June 2011, on the restriction of the use of certain hazardous substances in electrical and electronic equipment, whose Art. 2.4 reads 'This Directive does not apply to: (a) equipment which is necessary for the protection of the essential interests of the security of Member States, including arms, munitions and war material intended for specifically military purposes; (...)'; an identical exemption is provided by Art. 2.3(a) of the Directive 2012/19/EU of 4 July 2012, on waste electrical and electronic equipment (WEEE).

> *entirely outside the scope of Community law* (...). As the Court has already held, the only articles in which the [EC] Treaty expressly provides for derogations applicable in situations which may affect public safety (...) deal with exceptional and clearly defined cases. It cannot be inferred that the Treaty contains an inherent general exception excluding all measures taken for reasons of public security from the scope of Community law.[59]

Indeed, even in similar circumstances, Member States are anyway bound to act in accordance with the duty of loyal cooperation (Article 4.3 of TEU – former Article 10 of TEC),[60] namely they 'shall facilitate the achievement of the Union's tasks and refrain from any measure which could jeopardise the attainment of the Union's objectives'. It seems that the same reasoning, *a fortiori*, should apply to the exemptions for 'national defence' or 'military purposes' provided for in relevant EU secondary legislation.

5.4 The Case of the 'Quirra' Test/Training Range in the Light of the European Court of Human Rights Case Law

In the previous part of this contribution it has been stressed the importance, for the purposes of the case under consideration, of the State's duty to provide timely the public with due information on situations of environmental contamination susceptible of affecting human health. Now, it is interesting to assess how the existence and the scope of such a duty have been acknowledged also under the European Convention for the Protection of Human Rights and Fundamental Freedoms, particularly by the ECtHR case law.

As it is well-known, the ECHR does not contain any provision expressly aimed at protecting the environment. Particularly, as it is pointed out in a *Manual on Human Rights and the Environment* adopted by the Council of Europe in 2005, '[t]he Convention is not designed to provide a general protection of the environment as such and does not expressly guarantee a right to a sound, quiet and healthy environment'.[61] However, this has not prevented the ECtHR from deriving environmental rights from other existing Convention's rights, such as the right to private and family life or the right to life. It is an increasing jurisprudence which is described by scholarship as 'greening' of human rights.[62] Among the environmental cases dealt with by the Court, firstly one should refer to the case *Guerra and others v. Italy.*[63] Here, the Court has held that the respondent State had not fulfilled

59 Judgment of 15 December 2009, in Case C-239/06, *Commission v. Italy*, §46.

60 Idem, §54.

61 See P. Birnie, A. Boyle and C. Redgwell, *International Law & Environment*, 3rd edn, New York, 2009, p. 275.

62 Birnie, Boyle and Redgwell, above n. 61. p. 282 et seq.

63 Judgment of 19 February 1998, case *Guerra and others v. Italy*.

its obligation to secure the applicants' right to respect for their private and family life, in breach of Article 8 of the Convention, *because it had not provided them with essential information that would have enabled them to assess the risks* they and their families might run, in continuing to live near a chemical factory producing fertilizers classified as high-risk and toxic substances.[64] Particularly, the Court has determined that such a duty to inform the public, about possible risks from exposure to a situation of serious environment's pollution, is to be considered as one of *positive obligations* inherent in effective respect for private or family life. Indeed, Article 8 of the Convention does not merely compel the State to abstain from arbitrarily interfering with the individual's private or family life: in addition to this primarily negative undertaking, the State has *a positive duty to take the necessary steps to ensure effective protection* of the same individual's right.[65] From this point of view, what seems worth emphasizing here is that, in the Court's opinion, the State has violated Article 8 *just by failing to provide the public with information on possible risks*: even though the Court has recalled that the factory had released large quantities of toxic substances and that the applicants, living near the factory, had been so exposed to these toxic emissions, it has not found it necessary to consider the possible serious effects on residents most affected. The mere fact that the latter had been exposed to a danger and the State had failed to inform them on possible risks has been considered enough by the Court to hold that the respondent State had not fulfilled its obligation to secure the applicants' right to respect for their private and family life, in breach of Article 8 of the Convention.[66]

It is worth stressing that, also in the *Quirra* case, after all one could acknowledge a State's failure to provide the public with essential information about the actual environmental situation. However, one must even point out two important differences in comparison with the *Guerra* case. Firstly, while in the latter the respondent State could have not been regarded as interfering with the applicants' private or family life, in the *Quirra* case the very interference by military authorities with the inhabitants' private or family life is exactly at issue, particularly through the negative effects of military activities on the environment and human health. In this context, it must recalled that, according to paragraph 2 of Article 8, the interference by the public authority may be justified if it is 'necessary in a democratic society in the interests of national security'. Thus, in principle, it might not

64 Judgment of 19 February 1998, case *Guerra and others v. Italy*, §60.

65 Idem, §58.

66 Idem, §60: 'The Court reiterates that *severe environmental pollution* may affect individuals' well-being and prevent them from enjoying their homes in such a way as to affect their private and family life adversely (...). In the instant case *the applicants waited, right up until the production of fertilisers ceased in 1994, for essential information that would have enabled them to assess the risks they and their families might run if they continued to live at Manfredonia*, a town particularly exposed to danger in the event of an accident at the factory. The Court holds, therefore, that the respondent State did not fulfil its obligation to secure the applicants' right to respect for their private and family life, in breach of Article 8 of the Convention' (emphasis added).

be excluded that even the kind of interference linked with military training activities be considered as 'necessary' in the interests of national security.

Leaving this last assessment aside, the second difference in comparison with the *Guerra* case is that, within the areas around the Quirra firing range, both the degree and the quality of the environmental pollution are still controversial. Indeed, even the secrecy that usually cover every military experimental activity certainly has not helped to assess the actual situation. Only recently the judiciary investigation has discovered, both inside the land area and in the sea area, several evident zones seriously affected and deteriorated, as a result of many years of firing training activities. But the most alarming kind of pollution that comes into play, in this case, is actually a silent and invisible danger: the risk from exposure to radioactive military waste, which potentially might have already affected many people and the ecosystem, even with long-term effects. Such a special kind of risk as well as the public perception of it, are indeed the core of the *Quirra* case. Moreover, possible causal links between the use of experimental weapons and cancer incidence among people living around the test range are still disputed and difficult to establish.

In sum, even the existence of a 'severe environmental pollution', as the Court's case law requires in order to determine an infringement of Article 8 of the ECHR, might be challenged in the case at issue. This explains why reference should be made also to other environmental cases, particularly those in which the risk complained before the Court was not so evident and undisputed, as it was in the *Guerra* case instead.[67]

From this perspective, the case of *L.C.B. v. The United Kingdom*[68] seems to be particularly relevant, because it is related to a situation of exposure to certain military atmospheric nuclear tests, carried out in the fifties at the UK Christmas Island. The applicant was a woman who claimed to have contracted leukaemia as a result of her father's exposure to radiation, while he was serving as a catering assistant at Christmas Island.[69] She also complained that the UK Government had not provided her parents with information regarding

67 The circumstance that, differently from other cases mentioned below, in the *Guerra* case the risk was undisputed, and public concerns over risk were objectively well-grounded, is stressed in C. Hilson, *Risk and the European Convention on Human Rights: Towards New Approach*, in *Cambridge Yearbook of European Studies*, 2008-2009. <https://www.reading.ac.uk/web/FILES/law/CYEL_Hilson_article_2_June_final.pdf>, p. 11 et seq.CYEL

68 Judgment of 9 June 1998, case *L.C.B. v. The United Kingdom.*

69 Idem, §10 et seq. Another case related with the nuclear tests at the UK Christmas Island is *McGinley and Egan v. UK*, Judgment of 9 June 1998, concerning two servicemen exposed to radiation during such experiments in the 1950s. Particularly, having later contracted diseases, they complained that the UK Government had failed to provide them with information which could have enabled them to establish a causal link, for the purposes of claiming a war pension, between their presence at Christmas Island and their subsequent health problems. This same case, together with the *L.C.B.* case, is discussed in Hilson, above n. 67, p. 6 et seq., under the ECtHR case-law where 'individuals have been exposed to a toxic risk in the past and have developed a resulting fear or anxiety concerning this exposure which leads them to link their current health problems with it. Typically, they want access to information on the exposure incident and subsequent monitoring data to try to prove a causal link between their illnesses and the exposure'.

the extent of her father's exposure to radiation and the related genetic risks, arguing that such information would have allowed monitoring her health from infancy, leading up to an earlier diagnosis and treatment of her illness.[70] It is worth stressing how this case deal with two issues which indeed come into play in the *Quirra* case too, namely the risk from exposure to radiation and the military nature of the radioactive source of contamination.

The Court has observed that, in principle, the question could have been considered in relation to Article 8, but it has chosen to examine it under Article 2 of the ECHR (Right to life). After considering that 'the first sentence of Article 2 § 1 enjoins the State not only to refrain from the intentional and unlawful taking of life, but also *to take appropriate steps to safeguard the lives of those within its jurisdiction*', the Court acknowledged that its task was to determine 'whether, *given the circumstances of the case*, the State did all that could have been required of it to prevent the applicant's life from being avoidably put at risk'.[71] What seems important to point out here is that, dealing with a very controversial situation, in view of the lack of certainty as regards the extent of the father's exposure, as well as the existence of a causal link between such an exposure and the applicant's illness, the Court has held that 'the State could only have been required of its own motion [to provide advice to her parents and to monitor her health] *if it had appeared likely at that time* that any such exposure of her father to radiation might have engendered *a real risk* to her health'.[72] In other words, in the Court's opinion, 'given the information available to the State *at the relevant time*', the State authorities were not under a duty to advise the applicant's parents or to monitor her health.[73] Moreover, the reference to the information available *at the relevant time,* clearly encompass even the available *scientific knowledge*. Indeed, after having declared itself not satisfied about the existence of *a causal link* between the exposure of a father to radiation and leukaemia in a child subsequently conceived, the Court has stated that '[it] could not reasonably hold, therefore, that *in the late 1960s*, the United Kingdom authorities could or should, on the basis of this unsubstantiated link, have taken action in respect of the applicant'.[74] Such reasoning seems to leave no room for any consideration in the light of the precautionary principle: it rather seems to require *scientific evidence of a causal link* between the exposure to radiation and the onset of the illness. Such scientific evidence could be very difficult to ascertain in a complicated situation like that of the *Quirra* case.

In the same vein, one could also refer to the case *Balmer-Schafroth and others v. Switzerland*, concerning the possible risks feared by some people living near a nuclear

70 Case *L.C.B. v. The United Kingdom*, §29.
71 Idem, §36, emphasis added.
72 Idem, §38.
73 Idem, §41, emphasis added.
74 Idem, §39, emphasis added.

power station.[75] Here, the Court has held that '(...) [the applicants had] failed to show that the operation of the [nuclear] power station exposed them personally to a danger that was *not only serious but also specific and, above all, imminent.* In the absence of such a finding, the effects on the population (...) therefore remained hypothetical'.[76] Apart from exemplifying a particularly restrictive approach adopted by the Court, which stresses the need that the risk be 'imminent', the *Balmer-Schafroth* case is also to be regarded as an example of a situation in which the public perception of the risk is indeed 'out of line' with the perception of the State's authorities.[77] Such a last remark seems to befit the *Quirra* case too, though in the latter the perception of the risk among the people potentially affected is not so widespread as could perhaps be expected.

Anyway, in the more recent Court's case law, a new promising approach seems to extend the scope of certain Convention's rights also to cases of not undisputed or non-visible environmental pollution. In this vein, for instance, one could refer to the case, *Taşkin and others v. Turkey,*[78] concerning the risks related with the use of sodium cyanide to extract gold in a mine. Here, the Government had contested the applicability of Article 8 of the ECHR, arguing that 'the risk referred to by the applicants was hypothetical, since it might materialise only in twenty to fifty years. This was not a serious and imminent risk'.[79] But this time, after reiterating that 'Article 8 applies to severe environmental pollution which may affect individuals' well-being and prevent them from enjoying their homes in such a way as to affect their private and family life adversely, without, however, seriously endangering their health',[80] the Court has added that

> [t]he same is true where the dangerous effects of an activity to which the individuals concerned are likely to be exposed have been *determined as part of an environmental impact assessment procedure* in such a way as to establish *a sufficiently close link* with private and family life for the purposes of Article 8 of the Convention. If this were not the case, the positive obligation on the State to take reasonable and appropriate measures to secure the applicant's rights under paragraph 1 of Article 8 would be set at naught.[81]

75 Judgment of 26 August 1997, case *Balmer-Schafroth and others v. Switzerland.*

76 Idem, §40, emphasis added.

77 On this point, see Hilson, above n. 67, p. 11 et seq. and p. 14 et seq.

78 Judgment of 10 November 2004, *Taşkin and others v. Turkey.*

79 Idem, §107.

80 Idem, §113.

81 Idem, §113, emphasis added. See Hilson, above n. 67, p. 17: 'This addition is crucial because otherwise, Article 8 would only be applicable in straightforward, visible, pollution cases and not, as in Taşkin, where there is a real and serious long-term, but less tangible, risk posed by an activity. There is, therefore, no need to show pollution or direct harm as such: all that is needed is to establish the existence of a risk as part of an environmental impact, risk assessment.'

In other words, the applicability of Article 8 is not in principle excluded, even in cases of *silent and invisible pollution*, like the situation of possible public's exposure to radiation of military source in the *Quirra* case, provided that the existence of the risk be determined as part of an environmental impact assessment procedure. Unfortunately, as regards the situation in the areas around the Quirra firing range, exactly such an environmental impact assessment procedure is still far from be concluded.

Another case which is worth quoting here is *Öneryildiz v. Turkey.*[82] Here, the applicant had lost nine members of his family, as a result of a methane explosion at a landfill site lying near his slum. Accordingly, the Court has considered the case under Article 2 of the ECHR. In this view, after reiterating that the first paragraph of Article 2 lays down a positive obligation on States to take appropriate steps to safeguard the lives of those within their jurisdiction, the Court has given a precise definition of this substantive aspect of Article 2 of the Convention, in the particular context of dangerous activities:

> [That positive obligation] entails above all *a primary duty on the State to put in place a legislative and administrative framework* designed to provide effective deterrence against threats to the right to life the lives of those within their jurisdiction. (…) This obligation indisputably applies in the particular context of dangerous activities, where, in addition, special emphasis must be placed on regulations geared to the special features of the activity in question, particularly with regard to the level of the potential risk to human lives. They must govern the licensing, setting up, operation, security and supervision of the activity and must make it compulsory for all those concerned to take practical measures to ensure the effective protection of citizens whose lives might be endangered by the inherent risks. *Among these preventive measures, particular emphasis should be placed on the public's right to information,* as established in the case law of the Convention institutions. (…) [T]his right, which has already been recognised under Article 8 (see *Guerra and Others* (…)) may also, in principle, be relied on for the protection of the right to life, particularly as this interpretation is supported by current developments in European standards.[83]

Considering the *Quirra* case also in the light of such a latter ruling, one should point out that, regrettably, the duty to put in place an appropriate 'legislative and administrative framework', as above required, has actually been complied with by Italy only in 2009, when the Italian Ministry of Defence issued a Decree on 'procedures for the management of materials and waste, as well as for the clean-up of sites and facilities directly intended for

82 Judgment of 30 November 2004, *Öneryildiz v. Turkey.*

83 Idem, §§89-90, emphasis added.

military defence and national security purposes'.[84] Before such a Ministerial Decree, it seems difficult to find out in the Italian legal system any regulations on military waste management expressly aimed at protecting the environment and public health from negative effects of military training activities.[85] Indeed, one could only refer to the Law of 24 December 1976, n. 898, concerning the military *servitudes*, which is only intended to provide for certain prohibitions inside and around the military zones, even during military training activities, and does not contain neither the word 'environment', nor the word 'health.

5.5 Concluding Remarks

The above examined ECtHR case law seems to confirm that, both under Article 2 and Article 8 of the ECHR, States have a duty to provide the public with information on situations of environmental pollution susceptible of affecting human health, a duty that clearly comes into play also in relation to the *Quirra* case. Particularly, what the Court itself has emphatically defined as 'the public's right to information' is intended as a preventive measure that States have to take in order to fulfil their positive obligations under those same Convention's rules. Nevertheless, for the purposes of the case at issue here, it is not easy to foresee how and to what extent such a 'public's right', in the Court's opinion, should be reconciled with other competing interests, such as national security or national defence, which inevitably come into play. It should be taken into account that the Court often acknowledges to the States a wider margin of appreciation in the choice of the preventive measures to adopt, when carrying out dangerous activities.

Anyway, the aforementioned *Öneryildiz* case seems to be interesting also from another point of view. Indeed, it is worth stressing how, in such a case, the Court has rejected the Turkish Government's argument that the applicant had knowingly chosen to set up his home illegally in the vicinity of a rubbish tip, despite the inherent risks.[86] On this point, the Court has emphasized that the authorities had made *no attempt to discourage* the

84 Ministry of Defence, Decree of 22 October 2009, *'Procedure per la gestione dei materiali e dei rifiuti e la bonifica dei siti e delle infrastrutture direttamente destinati alla difesa militare e alla sicurezza nazionale'*, Italian Official Gazette of 15 April 2010, n. 87. This Ministerial Decree was issued in accordance with the Legislative Decree of 3 April 2006, n. 156, relating to 'Norme in materia ambientale' (Italian Official Gazette of 14 April 2006, n. 96), which in turn is aimed at complying with a number of EU directives on environmental matters, such as, among others, the Environmental Liability Directive (2004/35/EC) and the Environmental Impact Assessment Directive (85/337/CEE, then repealed by Directive 2011/92/EU).

85 According to the answer to a parliamentary question, given on 27 December 2010 by a former Italian Minister of Defence (<www.senato.it/japp/bgt/showdoc/frame.jsp?tipodoc=Sindispr&leg=16&id=00518687&parse=si&toc=no>), since 2004 it would have been adopted a mysterious 'disciplinare ambientale', which 'in accordance with the applicable rules' would regulate the procedures for authorizing the activities carried out inside the Quirra firing range.

86 Case *Öneryildiz v. Turkey*, §103.

applicant from living in his slum, near the source of the risks: rather they had actually *tolerated or encouraged* the development of slum areas around the tip.[87] Such a remark seems to be relevant also in the assessment of the situation around the Quirra firing range, if one considers that, far from attempting to discourage the inhabitants from living near the area, till recent times the competent military authorities even used to authorize shepherds to herd the flock to pasture within the same zones where firing training activities were carried out.

Indeed, another aspect which is worth stressing here is that, as already said, the perception of the risk among the people living around the military area is not so widespread. Rather, one must acknowledge that the most part of local population seems to be not aware of possible risks, or even seems to have 'chosen' to disregard those possible risks. Many people are only interested in the job opportunities associated with the existence of the firing range. In the same vein, it is sadly significant to notice the shocking statement of a resident, reported by a local newspaper: 'It's better one dead of cancer more, than one job less.'[88]

In such a social and cultural context, it is not surprising that, in 2012, the Municipality of Villaputzu has filed an application against the Ministry of Defence lodged with the Regional Administrative Tribunal of Sardinia, demanding the annulment -and firstly the suspension- of an order issued by the Quirra firing range's Commander on 17 October 2011. By such an order, all stockbreeders and farmers had been forbidden to enter and pass through the land area of the firing range (as they used to do in the past, with the authorization of the same military authority). But what is worth mentioning is that the Administrative Tribunal has rejected the petition for suspension invoking the *precautionary principle* as 'widely applied by the [European] Court of Justice'.[89] Perhaps, it has been the first time that such a principle has been invoked in relation to the environmental situation existing around the Quirra firing range.

The same principle should inspire even the EU Commission, when considering possible situations of environmental degradation arising from military activities. Instead of invoking certain exemptions provided for in EU secondary legislation, as an excuse for not dealing with possible risks for the environment and human health from that source of pollution, the Commission should *at least* remind Member States that, in the light of the *precautionary principle*, as well as of their *duty of loyal cooperation*, even those exemptions, such as that for 'national defence' or 'military purposes', do not authorize them to act totally in contrast with the overarching interest of the protection of the environment. Indeed, Members States should be warned by the Commission that, even within the military field, they have to refrain from any conduct that could jeopardise the objectives of the EU/Euratom policy

87 Case *Öneryildiz v. Turkey*, §104 et seq.

88 See P. Mannironi, *L'isola ferita e il silenzio della politica*, in *La Nuova Sardegna*, 9 March 2014, p. 5.

89 Regional Administrative Tribunal of Sardinia, Order of 26 January 2012, *Municipality of Villaputzu v. Ministry of Defence*, Appl. No. 00025/2012.

on the environment, which ultimately also aims at protecting human health. Even in the area of the protection against pollution from military sources, Member States could not exercise their competence in total disregard of the 'shared competence' scheme which applies between themselves and the Union as regards the policy on the environment. Moreover, no rule under EU/Euratom Law could be interpreted as derogating their *erga omnes* customary duty to prevent, reduce and control harm even to their own domestic environment. This duty includes the obligation to inform the public, without delay, on possible risks arising from exposure to a contaminated environment, which seems exactly the situation of the territory around the Quirra military test/training range.

6 Euratom and the Duty of Care for the Environment

Erik V. Koppe

6.1 Introduction

The European Atomic Energy Community (Euratom) was founded in 1957 by the six original founders of the current European Union pursuant to the Treaty Establishing the European Atomic Energy Community (Euratom Treaty).[1] According to the treaty's preamble, the States Parties to the Euratom Treaty wished to enhance and develop the peaceful use of nuclear energy because it would contribute to the prosperity of the peoples of the member states. Euratom was therefore the regional equivalent of the International Atomic Energy Agency (IAEA),[2] which was established in the same year in order 'to accelerate and enlarge the contribution of atomic energy to peace, health and prosperity throughout the world' (Article II of the Statute of the IAEA).

Unlike the Treaty Constituting the European Coal and Steel Community (ECSC),[3] which was concluded in 1952 for a period of 50 years, the Euratom Treaty was concluded for an unlimited period (Article 208 of the Euratom Treaty). It is still in force and provides the legal basis for rights and obligations of the States Parties to the Euratom Treaty as well as for Euratom itself.[4] Euratom has continued to exist as an independent international organization alongside the European Community and the European Union which were established in 1993 and in 2009. It is part of a single European Union legal order although its legal organs were merged with others after the 1965 merger treaty.[4]

The Euratom Treaty has been sparingly amended over the course of time and has proved to be a stable framework for the regulation of the peaceful use of nuclear energy

1 Consolidated Version of the Treaty Establishing the European Atomic Energy Community, Official Journal of the European Union, C 327, 26 October 2012 (2012/C 327/01) (EUR-Lex 11957A/TXT).

2 Statute of the International Atomic Energy Agency, signed on 23 October 1956, entered into force on 29 July 1957, through <www.iaea.org>. The IAEA was established following President Eisenhower's 'Atoms for Peace' speech before the General Assembly on 8 December 1953. T.N. Dupuy, G.M. Hammerman (Eds.), A Documentary History of Arms Control and Disarmament, T.N. Dupuy Associates, Dunn Loring, VA, 1973, pp. 358-364. See also <www.iaea.org>.

3 Treaty Constituting the European Coal and Steel Community, signed on 18 April 1951, entered into force on 23 July 1952, EUR-Lex 11951K/TXT.

4 Court of First Instance, Judgment of 6 May 2009 (*Outokumpu Oyj et al v. Commission of the European Communities*, Case T-122/04, ECLI:EU:T:2009:141, ECR 2009 II-0135, para. 55. See also Chapter 2 by Sousa Ferro in this volume.

in Euratom's member states. However, according to Trüe, 'some of its provisions appear outdated and others have been stretched in order to be able to address issues of major importance today, such as the safety and security of nuclear installations, their decommissioning, and the safe disposal of radioactive waste.'[5]

The commitment to the protection of the natural environment in the Euratom Treaty, is not included for example. Indeed, if you look at the terms of the Euratom Treaty, the case law of the European Court of Justice – as discussed elsewhere in this book – and the directives which have been promulgated, it appears that the main concern of the drafters is the protection of human health and security (see also for example Articles 2(b) and 30 Euratom Treaty. Powers are here attributed to Euratom to provide standards for the safety of workers and the general public). The lack of commitment to the protection of the natural environment may be 'insignificant in practice, as protection of the public may usually require protection of the environment'. Despite this it would certainly be 'desirable to improve legal clarity in this regard'.[6]

This chapter seeks to provide such clarity and discusses to what extent Euratom is bound by general rules of international environmental law. It discusses firstly to what extent Euratom is bound by general international law in the first place (Section 6.2) and subsequently discusses the development of international environmental law in general and the emergence of a customary duty of care for the environment in particular (Section 6.3). The chapter ends with a brief conclusion (Section 6.4).

6.2 Euratom and General International Law

Pursuant to Article 184 of the Euratom Treaty, Euratom has a separate legal personality. This means that it qualifies as a subject of law and that it can be subjected to legal rights and obligations. From its context it appears that Article 184 was intended to attribute legal personality under national law to Euratom and not under public international law. Indeed, Article 185 provides:

> In each of the Member States, [Euratom] shall enjoy the most extensive legal capacity accorded to legal persons under their laws; it may, in particular, acquire or dispose of movable and immovable property and may be a party to legal proceedings. To this end, [Euratom] shall be represented by the Commission.

5 C. Trüe, European Atomic Energy Community (EURATOM), in: Max Planck Encyclopedia of Public International Law, <www.mpepil.com>, para. 35.

6 Trüe, European Atomic Energy Community (EURATOM), para. 35.

It appears, however, that the founders of Euratom also intended the organization to have a legal personality under public international law. Firstly, Title III of the Euratom Treaty provides for an institutional structure of the organization, with a number of organs, which have the power to bind the member states. Secondly, Article 191 provides that Euratom shall enjoy those privileges and immunities in the territories of its member states as are necessary for the performance of its tasks. Thirdly, pursuant to Article 101, Euratom has the power to conclude agreements with third states and international organizations. These rights and powers, which Euratom has indeed exercised in practice,[7] can only be explained if it has a legal personality under public international law, not only vis-à-vis its member states, but also vis-à-vis third states.[8] It therefore qualifies as a subject of international law with the capacity to possess rights and duties under public international law.

According to the International Court of Justice, an international organization not only has those rights and obligations which have been specifically attributed to the organization under its constituent treaty or to which it has specifically consented by means of an agreement. Indeed 'international organizations are subjects of international law and, as such, are [also] bound by any obligations incumbent upon them under general rules of international law'.[9] Although the phrase 'general rules of international law' is ambiguous, it appears to refer to rules which are binding upon all subjects of international law.[10] As such, general rules of international law are most likely to materialize on the basis of custom, as evidenced by a general practice accepted as law.[11]

6.3 Euratom and the Duty of Care for the Environment

6.3.1 *Introduction*

It follows from the above that Euratom as an independent international organization with a separate legal personality is bound by rules of customary international law, including

7 Euratom has become party to multiple agreements, for example, including agreements concluded under the auspices of the IAEA on nuclear safety and, important for the topic under discussion, management of radioactive waste as will be discussed below.

8 Compare ICJ, Reparation for Injuries Suffered in the Service of the United Nations, Advisory Opinion, ICJ Reports 1949, p. 179, p. 185.

9 ICJ, Interpretation of the Agreement of 25 March 1951 between the WHO and Egypt, Advisory Opinion, ICJ Reports 1980, p. 73, para. 37.

10 See J.L. Kunz, General International Law and the Law of International Organizations, American Journal of International Law, Vol. 47, 1953, pp. 456-457.

11 This is confirmed by Schermers and Blokker, who write that international organizations are bound by the treaties they conclude, by custom, by general principles of law, and sometimes even by decisions of other international organizations. H.G. Schermers, N.M. Blokker, International Institutional Law, Fifth Revised Edition, Martinus Nijhoff Publishers, Leiden, 2011, paras. 1572-1581.

the rules of customary international environmental law. This section briefly discusses the development of international environmental law (Section 3.2), the meaning and scope of general principles of international environmental law (Section 3.3), the prohibition to cause significant transboundary pollution (Section 3.4) and, finally, the arguable duty of care for the environment as such (Section 3.5).

6.3.2 *The Development of International Environmental Law*

International environment law is a relatively new area of public international law which comprises of 'those substantive, procedural and institutional rules of international law that have as their primary objective the protection of the environment'.[12] There is no all-inclusive definition of 'the environment' as such, but the concept is believed to encompass a broad range of areas and issues, such as the protection of the marine environment, international water courses, the atmosphere, endangered species, natural resources and biodiversity.[13] In any case, 'the environment is not an abstraction but represents the living space, the quality of life and the very health of human beings, including generations unborn.'[14]

Although the importance of environmental protection was already recognized as early as 1867, when France and the United Kingdom concluded a treaty relating to fisheries,[15] modern international environmental law (or rather international law relating to the environment) generally developed after the United Nations Conference on the Human Environment in Stockholm in 1972. The impetus for this development was primarily given by the (Stockholm) Declaration which was adopted at the final day of the conference and which was intended 'to inspire and guide the peoples of the world in the preservation and enhancement of the human environment'.[16] The Stockholm Declaration contains 26 principles which link protection of the environment with human rights, elaborate on the rationale for environmental protection, provide means for environmental protection, and which embed the principles in relation to environmental protection in the general corpus

12 P. Sands and J. Peel et al., Principles of International Environmental Law, Third Edition, Cambridge University Press, Cambridge, 2012, p. 13.

13 See P. Birnie, A. Boyle, C. Redgwell, International Law & the Environment, Third Edition, Oxford University Press, Oxford, 2009, p. 6.

14 ICJ, Legality of the Threat or Use of Nuclear Weapons, Advisory Opinion of 8 July 1996, ICJ Reports 1996 (Nuclear Weapons Advisory Opinion), para. 29.

15 M. Fitzmaurice, International Protection of the Environment, 293 Recueil des Cours (2001), p. 28.

16 A/CONF.48/14/Rev.1, Report of the United Nations Conference on the Human Environment, Stockholm, 5-16 June 1972, Declaration of the United Nations Conference on the Human Environment, United Nations, New York, 1973, pp. 3-5, preamble. The Report was taken note of with satisfaction by the General Assembly by A/Res/2994 (XXVII), adopted on 15 December 1972, by 112 to 0, with 10 abstentions; United Nations Conference on the Human Environment.

of public international law.[17] The Stockholm Declaration not only established national departments for the protection of the environment all over the world, but also the establishment of the United Nations Environment Programme (UNEP).[18] by the General Assembly in 1972. Its purpose being '[t]o promote international cooperation in the field of the environment and to recommend, as appropriate, policies to this end'.[19]

6.3.3 General Principles of International Environmental Law

Modern international environmental law is founded on a number of general principles, such as the principles of prevention, the precautionary principle,[20] and the principle of good neighborliness[21] Some of these principle have been laid down in the 1972 Stockholm Declaration and in its successors, such as the 1982 World Charter for Nature[22] and the 1992 Rio Declaration.[23] Further, some of these principles have been recognized in case law from the International Court of Justice, the International Tribunal for the Law of the Seas, and arbitral tribunals.[24]

There appears to be no consensus as to the status of these principles under public international law despite their fundamental character. According to some, the principles of international environmental law qualify as general principles of international law and

17 See Sands and Peel et al., Principles of International Environmental Law, p. 24.

18 C. Redgwell, International Environmental Law, in: M. Evans (Ed.), International Law, Third Edition, Oxford University Press, Oxford, 2010, p. 689.

19 A/RES/2997 (XXVII) of 15 December 1972; Institutional and financial arrangements for International Environmental cooperation.

20 See, for example, Separate Opinion of Judge Cançado Trindade in relation to the Case Concerning Pulp Mills on the River Uruguay (*Argentina v. Uruguay*), Judgment of the ICJ of 20 April 2010 (*Pulp Mills* case), paras. 52-96.

21 F. A. Nelissen, Van Stockholm, via Rio naar Johannesburg; Enige volkenrechtelijke beschouwingen over het beginsel van goed nabuurschap (From Stockholm, via Rio to Johannesburg; Some public international law observations about the principle of good neighbourliness), inaugural lecture Groningen University, 8 October 2002, T.M.C. Asser Institute, The Hague 2002, pp. 11-12. The principle of good neighborliness has also been recognized in the United Nations Charter (Preamble and Art. 74), 39 AJIL 1945, Supplement: Official Documents, p. 190.

22 A/Res/37/7, adopted on 28 October 1982, by 111 to 1, with 18 abstentions; World Charter for Nature; Annex: World Charter for Nature.

23 A/CONF.151/26/Rev.l (Vol. I), Report of the United Nations Conference on Environment and Development, Rio de Janeiro, 3-14 June 1992, Volume I, Resolutions Adopted by the Conference; Resolution 1, Adoption of Texts on Environment and Development; Annex I, Rio Declaration on Environment and Development, United Nations, New York, 1993, pp. 3-8.

24 See, for example, the Nuclear Weapons Advisory Opinion; ICJ, Gabčíkovo-Nagymaros Project (*Hungary/Slovakia*), Judgment of 25 September 1997, ICJ Reports 1997 (*Gabčíkovo-Nagymaros* case); the *Pulp Mills* case; International Tribunal for the Law of the Seas, Responsibilities and obligations of States with respect to activities in the Area, Advisory Opinion of 1 February 2011, ITLOS Reports 2011 (ITLOS Advisory Opinion); Award of the Arbitral Tribunal of 24 May 2005, in the Arbitration Regarding the Iron Rhine ('IJzeren Rijn') Railway between The Kingdom of Belgium and The Kingdom of the Netherlands (Iron Rhine Arbitral Award), <www.pca-cpa.org>.

therefore as general principles in the meaning of Article 38(1)(c) Statute.[25] According to others, the principles of international environmental law reflect customary law, emerging legal obligations, or some other normative value, and must be clearly distinguished from general principles of law.[26]

It appears that the qualification of these principles result from a discussion of legal theory, in particular the distinction between principles and rules.[27] In order to solve this conundrum, some authors, such as Beyerlin, have relied on the theory of Dworkin who distinguishes between rules, which are applicable in an all-or-nothing fashion, and principles, which must be taken into account when it comes to decision-making and which provide guidance to the addressees.[28] Also Birnie, Boyle and Redgwell compare the general principles of international environmental law with Dworkin's (constitutional) principles. They state that these principles 'lay down parameters which affect the way courts decide cases or how an international institution exercises its discretionary powers. They can set limits to provide guidance, or determine how conflicts between other rules or principles will be resolved. They may lack the supposedly harder edge of a "rule" or "obligation", but they should not be confused with non-binding' or emerging law.'[29] In other words: 'A rule answers the question 'what': a principle in effect answers the question 'why'.'[30]

It therefore appears that the character of a particular norm determines whether or not it qualifies as a rule or a principle. However, the character of the norm as such says nothing about the source of the norm in question. This is particularly relevant when the norm in question is unwritten, or, in the case of public international law, when the norm has not been laid down in a convention with a universal scope. How then do rules and principles materialize and how do you identify the general principles of (international environmental) law?

25 See, for example, Birnie, Boyle and Redgwell, who write that the principles which are included in the 1992 Rio Declaration qualify as general principles of international law in the meaning of Art. 38(1)(c) Statute. Birnie, Boyle, Redgwell, International Law & the Environment, pp. 26-28. They are less explicit in their subsequent discussion of 'Rights and Obligations of States Concerning Protection of the Environment' (Chapter 3, in particular pp. 106-114).

26 Sands Peel et al., Principles of International Environmental Law, p. 188.

27 Indeed, according to the Arbitral Tribunal in the *Iron Rhine* case, '[t]here is considerable debate as to what, within the field of environmental law, constitutes "rules" or "principles"; what is "soft law"; and which environmental treaty law or principles have contributed to the development of customary international law.' Iron Rhine Arbitral Award, para. 58.

28 U. Beyerlin, Different Types of Norms in International Environmental Law; Policies, Principles, and Rules, in: D. Bodansky, J. Brunnée, E. Hey (Eds.), The Oxford Handbook of International Environmental Law, Oxford University Press, Oxford, 2007. Also U. Beyerlin, T. Marauhn, International Environmental Law, Hart Publishing, Oxford, 2011, pp. 37-38.

29 Birnie, Boyle, Redgwell, International Law & the Environment, p. 28.

30 G. Fitzmaurice, The General Principles of International Law; Considered from the standpoint of the rule of law, Recueil des Cours, Vol. 92, 1957-II (A.W. Sijthoff, Leiden, 1957) p. 7.

According to some, general principles of international (environmental) law obtain their 'authority and legitimacy' by the 'endorsement of states – *opinio juris* in other words'.[31] Although this view appears to emphasize the fact that general principles are part of positive law, the reference to endorsement by states and *opinio juris* carries an inherent risk of conflating the material source from which these principles are derived and may lead to qualifications such as customary principles or principles of customary international law.[32] Indeed, in the present author's view, custom, as reflected in a general practice accepted as law, will lead to specific rights and obligations which are applicable in an all-or-nothing fashion. In other words, custom creates rules and not general principles.

According to others, general principles should be 'inferred by way of induction and generalization from conventional and customary rules of international law'.[33] A similar view on the identification of legal principles was expressed by the eminent 20th century Dutch legal scholar Scholten. Scholten stated that (general) principles of law (*'rechtsbeginselen'*) reflect the *ratio legis* of laws or rules. They refer to the ethical/moral foundations of the rules and must be found within the system of rules as a whole by means of induction. Identification of legal principles by means of induction is an intellectual exercise and, according to Scholten, this is one of the most important tasks of legal scholarship.[34] Once discovered these legal principles may clarify other rules and regulations and the systematics of other areas of law and thus provide necessary (interpretative) tools for courts and tribunal for jurisprudence and the administration of justice.[35]

---

31 Birnie, Boyle, Redgwell, International Law & the Environment, p. 28. See also Beyerlin and Marauhn who state that only 'few principles, if any, have passed the threshold for having gained the status of a customary international norm' because '[p]roving that a state was directed by a certain principle when it took a decision on a particular action may be difficult in practice'. Beyerlin, Marauhn, International Environmental Law, p. 38.

32 According to Judge Cançado Trindade it is even 'conceptually flawed' to refer to customary principles of international law. Separate Opinion of Judge Cançado Trindade in relation to the Case Concerning Pulp Mills on the River Uruguay (*Argentina v. Uruguay*), Judgment of 20 April 2010, para. 17. According to Schwarzenberger, '[i]f the principle in question is said to be based on international customary law, this terminology easily becomes misleading. It is then an abbreviation for stating more accurately that the principle is an abstraction, and nothing but an abstraction, from a number of rules which themselves bear the hall-mark of this law-creating process.' G. Schwarzenberger, The fundamental principles of international law, Recueil des Cours, Vol. 87, 1955-1 (A.W. Sijthoff, Leiden, 1955) p. 207.

33 A. Cassese, International Law in a Divided World, Clarendon Press, Oxford, 1986, p. 174. Similarly, A. Cassese, International Law, Oxford University Press, Oxford, 2001, p. 152.

34 Compare the present author's identification of the principle of ambituity (derived from the Latin word ambitus, which means environment) by means of induction as a fundamental principle of the law of armed conflict. E.V. Koppe, The Principle of Ambituity and the Prohibition against Excessive Collateral Damage to the Environment during Armed Conflict, Nordic Journal of International Law, Vol. 82, 2013. The principle of ambituity safeguards the intrinsic value of the natural environment and is equivalent to the principle of humanity.

35 G.J. Scholten, Algemeen Deel; Mr. C. Asser's Handleiding to de beoefening van het Nederlands Burgerlijk Recht, W.E.J. Tjeenk Willink, Zwolle, 1974, pp. 62-65.

Identification of the general principles of (specific areas of) international law which underlie the corpus of positive public international law (by means of induction) would be desirable. It would provide guidance for the further development of international law,[36] to courts and tribunals for the interpretation and application of public international law[37] and it would clarify the interrelationship between various areas of public international law (unity within diversity). The International Law Commission appears to be ideally positioned to take on this responsibility as I, the author have argued previously.[38]

In view of the above, it is appealing to further qualify general principles of international environmental law as general principles of law in the meaning of Article 38(1)(c) Statute and/or as an autonomous source of international law. Although Article 38(1)(c) Statute was originally intended to encompass general rules of law *in foro domestico, i.e.* principles of domestic law that are common to all legal systems, such as the principle of good faith,[39] there appears to be increasing recognition that it also includes general principles of inter-

36 Compare for example the outcomes of a research carried out in the Netherlands upon the request of the government to re-codify the principles of environmental law: Ch.W. Backes, C.J. Bastmeijer, A.A. Freriks, R.A.J. van Gestel, J.M. Verschuuren, Codificatie van milieurechtelijke beginselen in de Wet Milieubeheer (Codification of principles of environmental law in the Environmental Management Act), Boom Juridische uitgevers, Den Haag, 2002.

37 Similarly Birnie, Boyle and Redgwell, who write that general principles of international environmental law have a guiding character and 'influence the interpretation and application of customary law'. Birnie, Boyle, Redgwell, International Law & the Environment, p. 28.

38 Koppe, The Principle of Ambituity and the Prohibition against Excessive Collateral Damage to the Environment during Armed Conflict, pp. 63-64.

39 A. Pellet, Article 38, in: A. Zimmermann et al., The Statute of the International Court of Justice; A Commentary, Oxford University Press, Oxford, 2006, pp. 766–773.

national law,[40] including principles of specific areas of international law, such as international environmental law.[41]

6.3.4 *The Prohibition to Cause Significant Transboundary Pollution*

There appears to be consensus with respect to at least one general principle of international environmental law, namely the principle of prevention / good neighborliness, which is also known under the maxim *sic utere tuo ut alienum non laedas.*[42] This general principle is reflected in, or underlies the customary rule which prohibits states to cause significant transboundary pollution. This prohibition was first recognized by an arbitral tribunal in the 1930s,[43] and requires states 'to ensure that activities within their jurisdiction and control respect the environment of other States or of areas beyond national control is now part of the corpus of international law relating to the environment.'[44]

---

40 Mosler identified general principles of international legal relations as one category of general principles covered by Art. 38(1)(c) Statute. These were 'principles resulting from the sovereign equality of States (jurisdiction, immunities, self-determination), principles connected with treaties, elementary considerations of humanity, and 'generally accepted principles' related to the law of the seas, diplomatic protection and armed conflict. H. Mosler, General Principles of Law, in: R. Bernhardt (ed.), Encyclopedia of Public International Law; Volume Two; East African Community to Italy-United States Air Transport Arbitration (1965), North-Holland Publishing Company, Amsterdam, 1995, pp. 513-514, 522-524. According to Judge Gaja, the ICJ often refers to principles of international law or general principles which have no parallel in domestic law. The identification of such principles by the ICJ 'whether or not it finds a parallel in municipal systems, is only rarely accompanied by an adequate demonstration of its existence in international law. G. Gaja, General Principles of Law, in: Max Planck Encyclopedia of Public International Law, through <www.mpepil.com>, paras. 17-20. Gaja refers in this context to the ICJ's recognition of the principle of *uti possidetis iuris* in the Frontier Dispute (Burkina Faso v. Republic of Mali), Judgment of 22 December 1986, para. 24. Tunkin even argued that Art. 38(1)(c) Statute ICJ only referred to general principles of international law. He refers to the fact that in 1945, the drafters of the Statute of the ICJ amended the chapeau of Art. 38 of the Statute of the Permanent Court of International Justice (PCIJ) by including the phrase 'whose function is to decide in accordance with international law such disputes as are submitted to it'. This amendment made Tunkin come to the conclusion that the scope of Art. 38(1)(c) would as from 1945 be limited to general principles of international law. G.I. Tunkin, 'General Principles of Law in International Law', in: R. Marcic (eds.), Internationale Festschrift für Alfred Verdross: zum 80. Geburtstag, Fink, München, 1971, p. 525. See also Cassese (apparently inspired by early writings of Ago), International Law in a Divided World, p. 174, 126.

41 Indeed, according to Judge Cançado Trindade, '[t]here are, in fact, general principles of law proper to international law in general, and there are principles of Law proper to some domains of International Law, such as, *inter alia*, International Environmental Law'. Separate Opinion of Judge Cançado Trindade in relation to the Case Concerning Pulp Mills on the River Uruguay (*Argentina v. Uruguay*), Judgment of 20 April 2010, para. 48. See also Cassese, International Law, pp. 152, 158.

42 Birnie, Boyle, Redgwell, International Law & the Environment, p. 137.

43 Trail Smelter Arbitration (*US v. Canada*); 16 April 1938, 11 March 1941, in: H. Lauterpacht (Ed.), Annual Digest and Reports of Public International Law Cases; Being a Selection from the Decisions of International and National Courts and Tribunals given during the Years 1938-1940, (also published as International Law Reports, Vol. 9), Butterworth & Co. (Publishers), London, 1942, Case No. 104 (pp. 315-333), p. 317.

44 Nuclear Weapons Advisory Opinion, para. 29; ICJ, Gabčíkovo-Nagymaros Project (Hungary/Slovakia), Judgment of 25 September 1997, ICJ Reports 1997 (*Gabčíkovo-Nagymaros* case), para. 53.

According to the ICJ in *Pulp Mills*, the required standard of behavior to prevent significant transboundary pollution is due diligence, which requires states to:

> undertake an environmental impact assessment where there is a risk that the proposed industrial activity may have a significant adverse impact in a transboundary context, in particular, on a shared resource. Moreover, due diligence, and the duty of vigilance and prevention which it implies, would not be considered to have been exercised, if a party planning works liable to affect the régime of the river or the quality of its waters did not undertake an environmental impact assessment on the potential effects of such works. (...) The Court also considers that an environmental impact assessment must be conducted prior to the implementation of a project. Moreover, once operations have started and, where necessary, throughout the life of the project, continuous monitoring of its effects on the environment shall be undertaken.[45]

The obligation to prevent significant transboundary harm appears to be closely related to the general obligation of states 'not to allow knowingly its territory to be used for acts contrary to the rights of other states', which was recognized by the International Court of Justice in the 1949 Corfu Channel Case.[46] This obligation is intended to protect the interests of a particular state, including the interests of its citizens and their property and seems to imply a general duty of care of states towards other states, similar to the general duties of care recognized in the theory of tort liability.[47]

In view of their close relationship and underlying rationale, it is arguable that the above-mentioned norms can be amalgamated into one single norm: a general and customary duty of care towards other states, more specifically a general duty of care for the environ-

---

45 ICJ, Case concerning Pulp Mills on the River Uruguay (*Argentina v. Uruguay*), Judgment of 20 April 2010 (*Pulp Mills* case), paras. 204-205. See also Principle 17 of the Rio Declaration. The principle of prevention has been further elaborated by the International Law Commission (ILC) in its 2001 Draft Articles on Transboundary Pollution, International Law Commission, Draft articles on prevention of Transboundary Harm from Hazardous Activities, with commentaries, 2001, Report of the International Law Commission on the work of its fifty-third session (A/56/10). These Draft Articles, provide, in short, that the state of origin must take all appropriate measures to prevent significant transboundary harm or minimize the risk thereof (Art. 3). For that purpose, states must, among other things, cooperate in good faith, seek assistance, if necessary, of international organizations (Art. 4), and take legislative and administrative measures to implement these articles (Art. 5). Further each state shall make sure that private parties do not carry out hazardous activities in their jurisdictions without the state's prior authorization. Such authorization procedure requires a proper environmental impact assessment as well as notification of the states involved, informing them about the outcomes, and consultation with them, with a view to achieve acceptable solutions in order to prevent harm or minimize the risk thereof (Arts. 6-9).

46 ICJ, *Corfu Channel* case, Judgment of 9 April 1949, ICJ Reports 1949 (*Corfu Channel* case), p. 22.

47 See M.A. Fitzmaurice, The Corfu Channel Case and the Development of International Law, in: N. Ando et al. (Eds.), Liber Amicorum Judge Shigeru Oda, Kluwer Law International, The Hague, 2002, pp. 132, 137-139. Fitzmaurice relies, among other things, on the dissenting opinion of Judge Azevedo in this regard.

ment in other states and for the environment in areas beyond national jurisdiction. As such, this obligation is intended to protect the interests of all states, or rather all mankind.[48]

6.3.5 *The Duty of Care for the Environment*

Furthermore, it is arguable that this duty of care is not limited to the environment in other states and to the environment in areas beyond national jurisdiction, but also extends to the environment – both the human environment and the environment as such – *within* a state's own jurisdiction.[49] Indeed, according to Sands (and Peel) the 'preventive principle' 'seeks to minimize environmental damage as an objective in itself' and may require states 'to prevent not only transboundary harm, but also damage to the environment *within its own jurisdiction*, including by means of appropriate regulatory, administrative and other measures.'[50] Similarly, Viñuales pertains that there are a number of reasons suggesting that 'customary international law requires States to prevent harm to the environment within their territory or jurisdiction even when no external impact can be reasonably foreseen.'[51]

Although this view is rejected by others,[52] there appears to be convincing evidence to identify a general duty of care for the environment under public international law. Such

---

48 *Gabčíkovo-Nagymaros* case, para. 53where the Court quotes the International Law Commission.

49 See also K.D. Jesse, E.V. Koppe, Business Enterprises and the Environment; Corporate Environmental Responsibility, Dovenschmidt Quarterly, December 2013, pp. 178-181.

50 Sands Peel et al., Principles of International Environmental Law, p. 201 (emphasis added). See also P. Sands, Principles of international environmental law; Vol. I: Frameworks, standards and implementation, Manchester University Press, 1995, p. 195 (with emphasis). This view was based, among other things, on a view expressed in 1986 by Singh, then President of the ICJ, in relation to a number of principles which had been drafted by the Experts Group on Environmental Law of the World Commission on Environment and Development. He wrote in the foreword of this report: 'The general principles concerning natural resources and environmental interferences recommended by the WCED Legal Experts Group (Articles 1-8) have in common that they are applicable to *all* instances of the use of a natural resource or of an environmental interference in *any* part of the world. It follows that these general principles do not merely apply in *areas beyond the limits of national jurisdiction* or in *the transboundary context*, but also in *the entirely domestic domain*, an area which, according to traditional international law, is subject to the exclusive jurisdiction of State. Such principles accordingly purport, as it were, to break open traditional international law concerning the use of natural resources or environmental interferences.' N. Singh, Foreword, in: R.D. Munro, J.G. Lammers (Eds.), Environmental Protection and Sustainable Development; Legal Principles and Recommendations, Graham & Trotman / Martinus Nijhoff, London / Dordrecht / Boston, 1987, pp. xi-xii.

51 J.E. Viñuales, Principle 2: Prevention, in: J.E. Viñuales (Ed.), The Rio Declaration on Environment and Development; A Commentary, Oxford University Press, Oxford, 2015, p. 119.

52 Redgwell, for example, writes that 'notwithstanding significant growth in the body of general and particularized rules governing State conduct in respect of the environment, it remains the case that there is not yet any general customary or treaty law obligation on States to protect and preserve the environment *per se*. (...) [F]or a general obligation to protect and preserve the environment wherever situated, one looks in vain.' C. Redgwell, International Environmental Law, in: M.D. Evans (Ed.), International Law, Third Edition, Oxford University Press, Oxford, 2010, p. 688.

duty appears to be recognized by the International Court of Justice[53] and would follow from the general principle of prevention, as stated by Sands and Peel, as well as the principle of sustainable development, which requires states to 'reconcile economic development with protection of the environment'.[54] Although states indeed have 'the sovereign right to exploit their own resources pursuant to their own environmental and developmental policies',[55] such exploitation must be carried out with due regard for the environment.[56] Sohn had already observed in 1973 that:

> An over-broad interpretation of this sovereign right would be inconsistent with the rest of the [Stockholm] Declaration which emphasizes the fact that no part of the global environment can be separated from the rest and that it has to be preserved and improved for the benefit of all the people of both the present and future generations. No state can claim an absolute right to ruin its environment in order to obtain some transient benefits. It should think not only of the effect on other peoples but also about the future of its own people. It should not ruin the soil of its own country in order to get a few extra crops or to sell more wood or pulp. Destruction and depletion of irreplaceable resources are clearly condemned by the Declaration, even when there is no effect abroad, and a state cannot engage in such activities behind the shield of misconceived sovereignty.[57]

Likewise the Arbitral Tribunal in the *Iron Rhine* case observed:

> Environmental law and the law on development stand (...) as mutually reinforcing, integral concepts, which require that where development may cause significant harm to the environment there is a duty to prevent, or at least miti-

---

53 See the Court's general observation that its Order in the 1995 Nuclear Weapons Test Cases is 'without prejudice to the obligations of States to respect and protect the natural environment, obligations to which both New Zealand and France have in the present instance reaffirmed their commitment.' Request for an Examination of the Situation in Accordance with Paragraph 63 of the Court's Judgment of 20 December 1974 in the Nuclear Tests (*New Zealand v. France*) Case, ICJ Reports 1995, para. 64.

54 *Gabčíkovo-Nagymaros* case, para. 140.

55 Principle 21 Stockholm Declaration, first sentence and Principle 2 Rio Declaration, first sentence.

56 See also the Principles for Environmental Protection and Sustainable Development Adopted by the WCED Experts Group on Environmental Law (A/42/427; Development and International Co-operation: Environment; Report of the World Commission on Environment and Development). Principles 2 and 3 require states, among other things, to 'conserve and use the environment and natural resources for the benefit of present and future generations' and 'to maintain ecosystems and ecological processes the functioning of the biosphere'. See for a commentary Munro, Lammers (Eds.), Environmental Protection and Sustainable Development; Legal Principles and Recommendation.

57 L.B. Sohn,The Stockholm Declaration on the Human Environment, Harvard International Law Journal, Vol. 14, 1973 p. 492.

> gate, such harm (...). This duty, in the opinion of the Tribunal, has now become a principle of general international law.[58]

The *opinio juris* required for the existence of such duty of care, is clearly evidenced by or can be inferred from the World Charter for Nature, which was adopted by the General Assembly of the United Nations in 1982.[59] The World Charter 'proclaims (...) principles of conservation by which all human conduct affecting nature is to be guided and judged'. The first four general principles provide:

1. Nature shall be respected and its essential processes shall not be impaired.
2. The genetic viability on the earth shall not be compromised; the population levels of all life forms, wild and domesticated, must be at least sufficient for their survival, and to this end necessary habitats shall be safeguarded.
3. All areas of the earth, both land and sea shall be subject to these principles of conservation; special protection shall be given to unique areas, to representative samples of all the different types of ecosystems and to the habitats of rare or endangered species.
4. Ecosystems and organisms, as well as the land, marine and atmospheric resources that are utilized by man, shall be managed to achieve and maintain optimum sustainable productivity, but not in such a way as to endanger the integrity of those other ecosystems or species with which they coexist.

Further, the existence of a general duty of care for the environment as such is also evidenced by the worldwide proliferation of environmental legislation. States which have enacted environmental protection legislation include the United States, the United Kingdom, France, the Russian Federation, the People's Republic of China, Germany, Brazil, India, South Africa, Canada, Australia, and the Netherlands. There is an ever increasing number of states which incorporate environmental protection into human rights law by including

58 Iron Rhine Arbitral Award, para. 59. The Tribunal subsequently concluded that both Belgium and the Netherlands had to implement measures to protect the environment upon reactivation of the Iron Rhine railway. Although this dispute related to the exercise of a treaty guaranteed right of Belgium in the Netherlands in relation to a railway, and therefore did not relate to an economic activity in one state leading to transboundary harm in the territory of another state, the 'Tribunal [was] of the view that, by analogy, where a state exercises a right under international law within the territory of another state, considerations of environmental protection also apply.' Iron Rhine Arbitral Award, para 222. This appears to suggest that the protection of the environment as such was the underlying rationale of this decision.

59 A/Res/37/7, adopted on 28 October 1982; World Charter for Nature. On the normative value of General Assembly resolutions and their importance for the formation of customary international law, see the Nuclear Weapons Advisory Opinion, para. 70.

specific environmental rights in their national constitutions.[60] Similarly, a number of international human rights treaties and documents recognize environmental rights.[61]

Finally, the existence of a duty of care for the environment as such arguably follows from the recognition of such duty of care within the law of armed conflict.[62] When it is generally accepted that states have an obligation under customary international law to protect the intrinsic value of the environment in times of armed conflict, which qualifies as an exceptional situation and which triggers the applicability of a specific set of rules, then such duty of care would *a fortiori* apply in times of peace.

6.4 CONCLUSION

The consequence of establishing the existence of a general customary duty of care for the environment is that all states, as well as all international organizations, must show due diligence to prevent damage to the environment, not only in a transboundary context, but also within national jurisdictions. Such due diligence requires states and international organizations to take all appropriate measures to prevent damage to the environment, including carrying out environmental impact assessments on the potential effects of a project prior to implementation of such project as well as throughout the life of the project.

---

60 See D. Shelton, Developing substantive environmental rights, Journal of Human Rights and the Environment, Vol. 1, 2010, p. 89-120.

61 See Art. 12(1) and (2)(b) of the 1966 International Covenant on Social and Economic Rights; Art. 24 of the 1981 African Charter of Human and Peoples' Rights; Art. 11 of the 1988 Additional Protocol to the American Convention on Human Rights in the area of Economic, Social and Cultural Rights; and the Vienna Declaration on Human Rights (A/CONF.157/23 of 12 July 1993), para. 11. There is also an increasing recognition of the importance of a clean and safe environment within the framework of classic human rights, such as the right to life (see, e.g., European Court of Human Rights, 30 November 2004, *Öneryildiz v. Turkey*, Appl. No. 48939/99), the right to health, the right to property and the right to respect for private and family life (see, e.g., European Court of Human Rights, 10 January 2012, *Di Sarno and other v. Italy*, Appl. No. 30765/08)). Compare also the mandate of the Special Rapporteur on the implications for human rights of the environmentally sound management and disposal of hazardous substances and wastes, A/HRC/RES/21/17 of 27 September 2012 and A/HRC/RES/27/23 of 26 September 2014.

62 In 2005, the International Committee for the Red Cross (ICRC) concluded that pursuant to customary international humanitarian law '[m]ethods and means of warfare must be employed with due regard to the protection and preservation of the natural environment. In the conduct of military operations, all feasible precautions must be taken to avoid, and in any even to minimize, incidental damage to the environment. Lack of scientific certainty as to the effects on the environment of certain military operations does not absolve a party to the conflict from taking such precautions.' This rule applies within the framework of international armed conflict and arguably within the framework of non-international armed conflict (Henckaerts, Doswald-Beck 2005) Rule 44. As such, this obligation embodies a general duty of care for the environment in times of armed conflict. After all, due regard is merely the standard of conduct which must be observed, similar to the general standard of due diligence for the above-stated customary duty of care for other states. See generally on the customary duty of care for the environment, E.V. Koppe, The Use of Nuclear Weapons and the Protection of the Environment during International Armed Conflict, Hart Publishing, Oxford, 2008, pp. 248-256.

For Euratom this means that it will have to require, or perhaps remind its member states,[63] as a matter of public international law, to take all appropriate measures to prevent damage to the environment when it comes to nuclear energy projects, including the management and transport of radioactive waste, even if protection of the natural environment is not specified in the Euratom Treaty. After all, as discussed above, international organizations, as subjects of international law, are not only bound by any obligations under their constituent treaties, but also by obligations incumbent upon them by general rules of international law, including general rules of customary international law.

In addition to Euratom's obligation to take all appropriate measures to prevent damage to the environment under customary international law when it comes to regulation of the management and transport of radioactive waste, Euratom is also bound to consider protection of the environment under treaty law. Indeed, Euratom acceded to the 1997 Joint Convention on the Safety of Spent Fuel Management and on the Safety of Radioactive Waste Management (Joint Convention) on 4 October 2005.[64] One of the Joint Convention's objectives is to 'ensure that during all stages of spent fuel and radioactive waste management, there are effective defenses against potential hazards so that individuals, society and the environment are protected from harmful effects of ionizing radiation, now and in the future, in such a way that the needs and aspirations of the present generation are met without compromising the ability of future generations to meet their needs and aspirations' (Article 1(ii)). It subsequently provides for specific obligations of the parties to the convention to take appropriate steps to limit potential impacts of spent fuel and radioactive waste on the environment. As such, these obligations appear to confirm or, at least, run parallel to Euratom's obligations under customary international law to observe a duty of care for the environment.

It appears that Euratom has partly implemented its obligations by promulgating legislation relevant to the management of radioactive waste and spent fuel, including Council Directive 2011/70/Euratom of 19 July 2011 establishing a Community framework for the responsible and safe management of spent fuel and radioactive waste,[65] which is discussed elsewhere in this volume. The legal basis for promulgating this directive was found in

63 In Chapter 22 of Agenda 21 which was adopted at the United Nations Conference on Environment and Development in Rio de Janeiro in June 1992. Chapter 22 relates to 'Safe and Environmentally Sound Management of Radioactive Wastes' and intends to 'ensure that radioactive wastes are safely managed, transported, stored and disposed of, with a view to protecting human health and the environment, within a wider framework of an interactive and integrated approach to radioactive waste management and safety' (Section 22.3). Its program of action is directed to states, which should undertake activities to this purpose, 'in cooperation with relevant international organizations.

64 Joint Convention on the Safety of Spent Fuel Management and on the Safety of Radioactive Waste Management (Joint Convention) on 4 October 2005 (entry into force on 2 January 2006), INFCIRC/546, through <www.iaea.org>.

65 Council Directive 2011/70/Euratom of 19 July 2011 establishing a Community framework for the responsible and safe management of spent fuel and radioactive waste, OJ L 199/48, 2 August 2011, p. 48.

Chapter 3 of the Euratom Treaty, on health and safety, including Article 37 of the treaty which requires Member States to submit any plans for the disposal of radioactive waste to the Commission to establish to what extent these plans are 'liable to result in the radioactive contamination of the water, soil or airspace of another Member State'. According to the Court of Justice of the European Union, these articles 'form a coherent whole conferring upon the Commission powers of some considerable scope in order to protect the population and the environment against the risks of nuclear contamination' (preambular para. 5 of the Directive).

It is suggested that the overarching customary duty of care for the environment, which binds both states and international organizations, and which must be taken into consideration by Euratom in exercising its powers under the Euratom Treaty as a matter of public international law, will be a useful tool to further shape environmental policies in relation to the management of radioactive waste.

Council Directive 2011/70/EURATOM

of 19 July 2011 Establishing a Community Framework for the Responsible and Safe Management of Spent Fuel and Radioactive Waste

THE COUNCIL OF THE EUROPEAN UNION,

Having regard to the Treaty establishing the European Atomic Energy Community, and in particular Articles 31 and 32 thereof,

Having regard to the proposal from the European Commission, drawn up after obtaining the opinion of a group of persons appointed by the Scientific and Technical Committee from among scientific experts in the Member States,

Having regard to the opinion of the European Economic and Social Committee[1],

Having regard to the opinion of the European Parliament[2],

Whereas:

1. Article 2(b) of the Treaty establishing the European Atomic Energy Community ('Euratom Treaty') provides for the establishment of uniform safety standards to protect the health of workers and of the general public.
2. Article 30 of the Euratom Treaty provides for the establishment of basic standards for the protection of the health of workers and the general public against the dangers arising from ionising radiations.
3. Article 37 of the Euratom Treaty requires Member States to provide the Commission with general data relating to any plan for the disposal of radioactive waste.
4. Council Directive 96/29/Euratom[3] establishes basic safety standards for the protection of the health of workers and the general public against the dangers arising from ionising radiation. That Directive has been supplemented by more specific legislation.
5. As recognised by the Court of Justice of the European Union in its case-law, the provisions of Chapter 3 of the Euratom Treaty, on health and safety, form a coherent whole

1 Opinion of 4 May 2011 (not yet published in the Official Journal).
2 Opinion of 23 June 2011 (not yet published in the Official Journal).
3 OJ L 159, 29.06.1996, p. 1.

conferring upon the Commission powers of some considerable scope in order to protect the population and the environment against the risks of nuclear contamination.[4]

6. Council Decision 87/600/Euratom of 14 December 1987 on Community arrangements for the early exchange of information in the event of a radiological emergency[5] established a framework for notification and provision of information to be used by the Member States in order to protect the general public in case of a radiological emergency. Council Directive 89/618/Euratom of 27 November 1989 on informing the general public about health protection measures to be applied and steps to be taken in the event of a radiological emergency[6] imposed obligations on the Member States to inform the general public in the event of a radiological emergency.
7. Council Directive 2003/122/Euratom[7] provides for the control of high-activity sealed radioactive sources and orphan sources, including disused sources. In accordance with the Joint Convention on the Safety of Spent Fuel Management and on the Safety of Radioactive Waste Management ('the Joint Convention') and the International Atomic Energy Agency (IAEA) Code of Conduct on the Safety and Security of Radioactive Sources, and current industrial practices, disused sealed sources can be reused, recycled or disposed of. In many cases, this needs a return of the source or return of the equipment, including the source, to a supplier or a manufacturer, for requalification or processing.
8. Directive 2006/21/EC of the European Parliament and of the Council of 15 March 2006 on the management of waste from extractive industries[8] covers the management of waste from extractive industries which may be radioactive, but excluding such aspects as are specific to radioactivity, which are matters dealt with under the Euratom Treaty.
9. Council Directive 2006/117/Euratom[9] lays down a European Atomic Energy Community ('Community') system of supervision and control of transboundary shipments of radioactive waste and spent fuel. That Directive was supplemented by Commission Recommendation 2008/956/Euratom of 4 December 2008 on criteria for the export of radioactive waste and spent fuel to third countries.[10]
10. Council Directive 2009/71/Euratom of 25 June 2009 establishing a Community framework for the nuclear safety of nuclear installations[11] imposes obligations on the Member States to establish and maintain a national framework for nuclear safety. While that Directive concerns principally the nuclear safety of nuclear installations, it states

---

4 C-187/87 (1988 ECR p. 5013) and C-29/99 (2002 ECR p. I-11221).
5 OJ L 371, 30.12.1987, p. 76.
6 OJ L 357, 07.12.1989, p. 31.
7 OJ L 346, 31.12.2003, p. 57.
8 OJ L 102, 11.04.2006, p. 15.
9 OJ L 337, 05.12.2006, p. 21.
10 OJ L 338, 17.12.2008, p. 69.
11 OJ L 172, 02.07.2009, p. 18.

that it is also important to ensure the safe management of spent fuel and radioactive waste, including at storage and disposal facilities. Therefore, those facilities, addressed both in Directive 2009/71/Euratom and in this Directive, should not be subject to disproportionate or unnecessary obligations, especially as regards reporting.

11. Directive 2003/35/EC of the European Parliament and of the Council of 26 May 2003 providing for public participation in respect of the drawing up of certain plans and programmes relating to the environment[12] applies to certain plans and programmes within the scope of Directive 2001/42/EC of the European Parliament and of the Council of 27 June 2001 on the assessment of the effects of certain plans and programmes on the environment.[13]
12. Commission Recommendation 2006/851/Euratom of 24 October 2006 on the management of the financial resources for the decommissioning of nuclear installations, spent fuel and radioactive waste[14] focuses on the adequacy of funding, its financial security and its transparency in order to ensure that the funds are only used for the intended purposes.
13. Under the specific terms of accession of Lithuania, Slovakia and Bulgaria to the European Union, where certain nuclear power plants were subject to early shutdown, the Community has taken part in the raising of financial resources and provides financial support subject to certain conditions to various decommissioning projects, including management of radioactive waste and spent fuel.
14. The Joint Convention, concluded under the auspices of the IAEA, represents an incentive instrument which aims at achieving and maintaining a high level of safety worldwide in spent fuel and radioactive waste management through the enhancement of national measures and international cooperation.
15. Some Member States have already participated and intend to participate further in the US-Russian driven programme, called the Global Threat Reduction Initiative, by shipping the spent fuel of research reactors to the United States of America and to the Russian Federation.
16. In 2006, the IAEA updated the structure of standards and published the Fundamental Safety Principles, which were jointly sponsored by the Community, the Organisation for Economic Cooperation and Development/Nuclear Energy Agency and other international organisations. Applying the Fundamental Safety Principles will facilitate the application of international safety standards and will make for greater consistency between the arrangements of different states.

12 OJ L 156, 25.06.2003, p. 17.

13 OJ L 197, 21.07.2001, p. 30.

14 OJ L 330, 28.11.2006, p. 31.

17. Following the Council's invitation to set up a High Level Group at EU level, as recorded in its Conclusions of 8 May 2007 on Nuclear Safety and Safe Management of Spent Nuclear Fuel and Radioactive Waste, the European Nuclear Safety Regulators Group (ENSREG) was set up by Commission Decision 2007/530/Euratom of 17 July 2007 on establishing the European High Level Group on Nuclear Safety and Waste Management[15] to contribute to the achievement of the Community objectives in the field of spent fuel and radioactive waste management. The conclusions and recommendations of ENSREG were reflected in the Council Resolution of 16 December 2008 on Spent Fuel and Radioactive Waste Management and the Council Conclusions of 10 November 2009 on the report by the European Nuclear Safety Regulators Group.
18. The European Parliament adopted on 10 May 2007 a Resolution 'Assessing Euratom – 50 Years of European nuclear energy policy' where it called for harmonised standards for radioactive waste management and invited the Commission to review the relevant drafts of its legislative proposal and submit a new proposal for a directive on radioactive waste management.
19. While each Member State remains free to define its energy mix, all Member States generate radioactive waste from power generation or in the course of industrial, agricultural, medical and research activities, or through decommissioning of nuclear facilities or in situations of remediation and interventions.
20. The operation of nuclear reactors generates spent fuel. Each Member State remains free to define its fuel cycle policy. The spent fuel can be regarded either as a valuable resource that may be reprocessed or as radioactive waste that is destined for direct disposal. Whatever option is chosen, the disposal of high-level waste, separated at reprocessing, or of spent fuel regarded as waste should be considered.
21. Radioactive waste, including spent fuel considered as waste, requires containment and isolation from humans and the living environment over the long term. Its specific nature, namely that it contains radionuclides, requires arrangements to protect human health and the environment against dangers arising from ionising radiation, including disposal in appropriate facilities as the end location point. The storage of radioactive waste, including long-term storage, is an interim solution, but not an alternative to disposal.
22. A national radioactive waste classification scheme should support those arrangements, taking fully into account the specific types and properties of radioactive waste.
23. The typical disposal concept for low and intermediate-level waste is near-surface disposal. It is broadly accepted at the technical level that, at this time, deep geological disposal represents the safest and most sustainable option as the end point of the management of high-level waste and spent fuel considered as waste. Member States,

15 OJ L 195, 17.07.2007, p. 44.

while retaining responsibility for their respective policies in respect of the management of their spent fuel and low, intermediate or high-level radioactive waste, should include planning and implementation of disposal options in their national policies. Since the implementation and development of a disposal facility will take place over many decades, many programmes recognise the necessity of remaining flexible and adaptable, e.g. in order to incorporate new knowledge about site conditions or the possible evolution of the disposal system. The activities conducted under the Implementing Geological Disposal of Radioactive Waste Technology Platform (IGD-TP) could facilitate access to expertise and technology in this respect. To that end, reversibility and retrievability as operating and design criteria may be used to guide the technical development of a disposal system. However, those criteria should not be a substitute for a well designed disposal facility that has a defensible basis for closure. A compromise is needed as the management of radioactive waste and spent fuel is based on state-of-the-art science and technology.

24. It should be an ethical obligation of each Member State to avoid any undue burden on future generations in respect of spent fuel and radioactive waste including any radioactive waste expected from decommissioning of existing nuclear installations. Through the implementation of this Directive Member States will have demonstrated that they have taken reasonable steps to ensure that that objective is met.
25. The ultimate responsibility of Member States for the safety of spent fuel and radioactive waste management is a fundamental principle reaffirmed by the Joint Convention. That principle of national responsibility, as well as the principle of prime responsibility of the licence holder for the safety of spent fuel and radioactive waste management under the supervision of its competent regulatory authority, should be enhanced and the role and independence of the competent regulatory authority should be reinforced by this Directive.
26. It is understood that the utilisation of radioactive sources by a competent regulatory authority for the purpose of carrying out its regulatory tasks does not affect its independence.
27. Member States should ensure that adequate funding is available for the management of spent fuel and radioactive waste.
28. Member States should establish national programmes to ensure the transposition of political decisions into clear provisions for the timely implementation of all steps of spent fuel and radioactive waste management from generation to disposal. It should be possible for such national programmes to be in the form of a single reference document or a set of documents.
29. It is understood that national arrangements for the safety of spent fuel and radioactive waste management will be applied through some form of legal, regulatory or organisa-

tional instrument, the choice of which rests within the competence of the Member States.

30. The different steps in spent fuel and radioactive waste management are closely interrelated. Decisions taken in one individual step may affect a subsequent step. Therefore such interdependencies should be taken into account when developing national programmes.
31. Transparency is important in the management of spent fuel and radioactive waste. Transparency should be provided by ensuring effective public information and opportunities for all stakeholders concerned, including local authorities and the public, to participate in the decision-making processes in accordance with national and international obligations.
32. Cooperation between Member States and at an international level could facilitate and accelerate decision-making through access to expertise and technology.
33. Some Member States consider that the sharing of facilities for spent fuel and radioactive waste management, including disposal facilities, is a potentially beneficial, safe and cost-effective option when based on an agreement between the Member States concerned.
34. The documentation of the decision-making process as it relates to safety should be commensurate with the levels of risk (graded approach) and should provide a basis for decisions related to the management of spent fuel and radioactive waste. This should enable the identification of areas of uncertainty on which attention needs to be focused in an assessment of safety. Safety decisions should be based on the findings of an assessment of safety and information on the robustness and reliability of that assessment and the assumptions made therein. The decision-making process should therefore be based on a collection of arguments and evidence that seek to demonstrate that the required standard of safety is achieved for a facility or activity related to the management of spent fuel and radioactive waste. In the particular case of a disposal facility, the documentation should improve understanding of those aspects influencing the safety of the disposal system, including natural (geological) and engineered barriers, and the expected development of the disposal system over time.
35. A Member State which has no spent fuel, no immediate prospect of having spent fuel and no present or planned activities related to spent fuel, would be under a disproportionate and unnecessary obligation if it had to transpose and implement the provisions of this Directive with regard to spent fuel. Therefore, such Member States should be exempted, for as long as they have not taken a decision to develop any activity related to nuclear fuel, from the obligation to transpose and implement the provisions related to spent fuel of this Directive.
36. A Treaty between the government of the Republic of Slovenia and the government of the Republic of Croatia on the regulation of the status and other legal relations

regarding investment, exploitation and decommissioning of the Krško nuclear power plant governs the co-ownership of a nuclear power plant. That Treaty provides for shared responsibility for the management and disposal of radioactive waste and spent fuel. Therefore an exemption to certain provisions of this Directive should be laid down in order not to hinder the full implementation of that bilateral Treaty.

37. While recognising that radiological and non-radiological hazards associated with spent fuel and radioactive waste should be taken into account in the national framework, this Directive does not cover non-radiological hazards, which fall under the Treaty on the Functioning of the European Union.
38. Maintenance and further development of competences and skills in the management of spent fuel and radioactive waste, as an essential element to ensure high levels of safety, should be based on learning through operational experience.
39. Scientific research and technological development supported by technical cooperation between actors may open horizons to improve the safe management of spent fuel and radioactive waste, as well as contribute to reducing the risk of the radiotoxicity of high-level waste.
40. Peer review could serve as an excellent means of building confidence and trust in the management of radioactive waste and spent fuel in the European Union, with the aim of developing and exchanging experience and ensuring high standards,

HAS ADOPTED THIS DIRECTIVE:

Chapter 1 Scope, Definitions and General Principles

Article 1 **Subject-matter**

1. This Directive establishes a Community framework for ensuring responsible and safe management of spent fuel and radioactive waste to avoid imposing undue burdens on future generations.
2. It ensures that Member States provide for appropriate national arrangements for a high level of safety in spent fuel and radioactive waste management to protect workers and the general public against the dangers arising from ionising radiation.
3. It ensures the provision of necessary public information and participation in relation to spent fuel and radioactive waste management while having due regard to security and proprietary information issues.
4. Without prejudice to Directive 96/29/Euratom, this Directive supplements the basic standards referred to in Article 30 of the Euratom Treaty as regards the safety of spent fuel and radioactive waste.

Article 2 **Scope**

1. This Directive shall apply to all stages of:
   a. spent fuel management when the spent fuel results from civilian activities;
   b. radioactive waste management, from generation to disposal, when the radioactive waste results from civilian activities.
2. This Directive shall not apply to:
   a. waste from extractive industries which may be radioactive and which falls within the scope of Directive 2006/21/EC;
   b. authorised releases.
3. Article 4(4) of this Directive shall not apply to:
   a. repatriation of disused sealed sources to a supplier or manufacturer;
   b. shipment of spent fuel of research reactors to a country where research reactor fuels are supplied or manufactured, taking into account applicable international agreements;
   c. the waste and spent fuel of the existing Krško nuclear power plant, when it concerns shipments between Slovenia and Croatia.
4. This Directive shall not affect the right of a Member State or an undertaking in that Member State to return radioactive waste after processing to its country of origin where:
   a. the radioactive waste is to be shipped to that Member State or undertaking for processing; or
   b. other material is to be shipped to that Member State or undertaking with the purpose of recovering the radioactive waste.

This Directive shall not affect the right of a Member State or an undertaking in that Member State to which spent fuel is to be shipped for treatment or reprocessing to return to its country of origin radioactive waste recovered from the treatment or reprocessing operation, or an agreed equivalent.

Article 3 **Definitions**

For the purpose of this Directive the following definitions shall apply:

1. 'closure' means the completion of all operations at some time after the emplacement of spent fuel or radioactive waste in a disposal facility, including the final engineering or other work required to bring the facility to a condition that will be safe in the long term;
2. 'competent regulatory authority' means an authority or a system of authorities designated in a Member State in the field of regulation of the safety of spent fuel or radioactive waste management as referred to in Article 6;
3. 'disposal' means the emplacement of spent fuel or radioactive waste in a facility without the intention of retrieval;

4. 'disposal facility' means any facility or installation the primary purpose of which is radioactive waste disposal;
5. 'licence' means any legal document granted under the jurisdiction of a Member State to carry out any activity related to the management of spent fuel or radioactive waste, or to confer responsibility for siting, design, construction, commissioning, operation, decommissioning or closure of a spent fuel management facility or of a radioactive waste management facility;
6. 'licence holder' means a legal or natural person having overall responsibility for any activity or facility related to the management of spent fuel or radioactive waste as specified in a licence;
7. 'radioactive waste' means radioactive material in gaseous, liquid or solid form for which no further use is foreseen or considered by the Member State or by a legal or natural person whose decision is accepted by the Member State, and which is regulated as radioactive waste by a competent regulatory authority under the legislative and regulatory framework of the Member State;
8. 'radioactive waste management' means all activities that relate to handling, pretreatment, treatment, conditioning, storage, or disposal of radioactive waste, excluding off-site transportation;
9. 'radioactive waste management facility' means any facility or installation the primary purpose of which is radioactive waste management;
10. 'reprocessing' means a process or operation, the purpose of which is to extract fissile and fertile materials from spent fuel for further use;
11. 'spent fuel' means nuclear fuel that has been irradiated in and permanently removed from a reactor core; spent fuel may either be considered as a usable resource that can be reprocessed or be destined for disposal if regarded as radioactive waste;
12. 'spent fuel management' means all activities that relate to the handling, storage, reprocessing, or disposal of spent fuel, excluding off-site transportation;
13. 'spent fuel management facility' means any facility or installation the primary purpose of which is spent fuel management;
14. 'storage' means the holding of spent fuel or of radioactive waste in a facility with the intention of retrieval.

Article 4 **General principles**

1. Member States shall establish and maintain national policies on spent fuel and radioactive waste management. Without prejudice to Article 2(3), each Member State shall have ultimate responsibility for management of the spent fuel and radioactive waste generated in it.
2. Where radioactive waste or spent fuel is shipped for processing or reprocessing to a Member State or a third country, the ultimate responsibility for the safe and responsible

disposal of those materials, including any waste as a by-product, shall remain with the Member State or third country from which the radioactive material was shipped.

3. National policies shall be based on all of the following principles:
   a. the generation of radioactive waste shall be kept to the minimum which is reasonably practicable, both in terms of activity and volume, by means of appropriate design measures and of operating and decommissioning practices, including the recycling and reuse of materials;
   b. the interdependencies between all steps in spent fuel and radioactive waste generation and management shall be taken into account;
   c. spent fuel and radioactive waste shall be safely managed, including in the long term with passive safety features;
   d. implementation of measures shall follow a graded approach;
   e. the costs for the management of spent fuel and radioactive waste shall be borne by those who generated those materials;
   f. an evidence-based and documented decision-making process shall be applied with regard to all stages of the management of spent fuel and radioactive waste.
4. Radioactive waste shall be disposed of in the Member State in which it was generated, unless at the time of shipment an agreement, taking into account the criteria established by the Commission in accordance with Article 16(2) of Directive 2006/117/Euratom, has entered into force between the Member State concerned and another Member State or a third country to use a disposal facility in one of them.

   Prior to a shipment to a third country, the exporting Member State shall inform the Commission of the content of any such agreement and take reasonable measures to be assured that:
   a. the country of destination has concluded an agreement with the Community covering spent fuel and radioactive waste management or is a party to the Joint Convention on the Safety of Spent Fuel Management and on the Safety of Radioactive Waste Management ('the Joint Convention');
   b. the country of destination has radioactive waste management and disposal programmes with objectives representing a high level of safety equivalent to those established by this Directive; and
   c. the disposal facility in the country of destination is authorised for the radioactive waste to be shipped, is operating prior to the shipment, and is managed in accordance with the requirements set down in the radioactive waste management and disposal programme of that country of destination.

Chapter 2 Obligations

Article 5 **National framework**

1. Member States shall establish and maintain a national legislative, regulatory and organisational framework ('national framework') for spent fuel and radioactive waste management that allocates responsibility and provides for coordination between relevant competent bodies. The national framework shall provide for all of the following:
   a. a national programme for the implementation of spent fuel and radioactive waste management policy;
   b. national arrangements for the safety of spent fuel and radioactive waste management. The determination of how those arrangements are to be adopted and through which instrument they are to be applied rests within the competence of the Member States;
   c. a system of licensing of spent fuel and radioactive waste management activities, facilities or both, including the prohibition of spent fuel or radioactive waste management activities, of the operation of a spent fuel or radioactive waste management facility without a licence or both and, if appropriate, prescribing conditions for further management of the activity, facility or both;
   d. a system of appropriate control, a management system, regulatory inspections, documentation and reporting obligations for radioactive waste and spent fuel management activities, facilities or both, including appropriate measures for the post-closure periods of disposal facilities;
   e. enforcement actions, including the suspension of activities and the modification, expiration or revocation of a licence together with requirements, if appropriate, for alternative solutions that lead to improved safety;
   f. the allocation of responsibility to the bodies involved in the different steps of spent fuel and radioactive waste management; in particular, the national framework shall give primary responsibility for the spent fuel and radioactive waste to their generators or, under specific circumstances, to a licence holder to whom this responsibility has been entrusted by competent bodies;
   g. national requirements for public information and participation;
   h. the financing scheme(s) for spent fuel and radioactive waste management in accordance with Article 9.
2. Member States shall ensure that the national framework is improved where appropriate, taking into account operating experience, insights gained from the decision-making process referred to in Article 4(3)(f), and the development of relevant technology and research.

Article 6 **Competent regulatory authority**

1. Each Member State shall establish and maintain a competent regulatory authority in the field of safety of spent fuel and radioactive waste management.
2. Member States shall ensure that the competent regulatory authority is functionally separate from any other body or organisation concerned with the promotion or utilisation of nuclear energy or radioactive material, including electricity production and radioisotope applications, or with the management of spent fuel and radioactive waste, in order to ensure effective independence from undue influence on its regulatory function.
3. Member States shall ensure that the competent regulatory authority is given the legal powers and human and financial resources necessary to fulfil its obligations in connection with the national framework as described in Article 5(1)(b), (c), (d) and (e).

Article 7 **Licence holders**

1. Member States shall ensure that the prime responsibility for the safety of spent fuel and radioactive waste management facilities and/or activities rest with the licence holder. That responsibility can not be delegated.
2. Member States shall ensure that the national framework in place require licence holders, under the regulatory control of the competent regulatory authority, to regularly assess, verify and continuously improve, as far as is reasonably achievable, the safety of the radioactive waste and spent fuel management facility or activity in a systematic and verifiable manner. This shall be achieved through an appropriate safety assessment, other arguments and evidence.
3. As part of the licensing of a facility or activity the safety demonstration shall cover the development and operation of an activity and the development, operation and decommissioning of a facility or closure of a disposal facility as well as the post-closure phase of a disposal facility. The extent of the safety demonstration shall be commensurate with the complexity of the operation and the magnitude of the hazards associated with the radioactive waste and spent fuel, and the facility or activity. The licensing process shall contribute to safety in the facility or activity during normal operating conditions, anticipated operational occurrences and design basis accidents. It shall provide the required assurance of safety in the facility or activity. Measures shall be in place to prevent accidents and mitigate the consequences of accidents, including verification of physical barriers and the licence holder's administrative protection procedures that would have to fail before workers and the general public would be significantly affected by ionising radiation. That approach shall identify and reduce uncertainties.
4. Member States shall ensure that the national framework require licence holders to establish and implement integrated management systems, including quality assurance,

which give due priority for overall management of spent fuel and radioactive waste to safety and are regularly verified by the competent regulatory authority.

5. Member States shall ensure that the national framework require licence holders to provide for and maintain adequate financial and human resources to fulfil their obligations with respect to the safety of spent fuel and radioactive waste management as laid down in paragraphs 1 to 4.

Article 8 **Expertise and skills**

Member States shall ensure that the national framework require all parties to make arrangements for education and training for their staff, as well as research and development activities to cover the needs of the national programme for spent fuel and radioactive waste management in order to obtain, maintain and to further develop necessary expertise and skills.

Article 9 **Financial resources**

Member States shall ensure that the national framework require that adequate financial resources be available when needed for the implementation of national programmes referred to in Article 11, especially for the management of spent fuel and radioactive waste, taking due account of the responsibility of spent fuel and radioactive waste generators.

Article 10 **Transparency**

1. Member States shall ensure that necessary information on the management of spent fuel and radioactive waste be made available to workers and the general public. This obligation includes ensuring that the competent regulatory authority inform the public in the fields of its competence. Information shall be made available to the public in accordance with national legislation and international obligations, provided that this does not jeopardise other interests such as, inter alia, security, recognised in national legislation or international obligations.
2. Member States shall ensure that the public be given the necessary opportunities to participate effectively in the decision-making process regarding spent fuel and radioactive waste management in accordance with national legislation and international obligations.

Article 11 **National programmes**

1. Each Member State shall ensure the implementation of its national programme for the management of spent fuel and radioactive waste ('national programme'), covering all types of spent fuel and radioactive waste under its jurisdiction and all stages of spent fuel and radioactive waste management from generation to disposal.

2. Each Member State shall regularly review and update its national programme, taking into account technical and scientific progress as appropriate as well as recommendations, lessons learned and good practices from peer reviews.

Article 12 **Contents of national programmes**

1. The national programmes shall set out how the Member States intend to implement their national policies referred to in Article 4 for the responsible and safe management of spent fuel and radioactive waste to secure the aims of this Directive, and shall include all of the following:
   a. the overall objectives of the Member State's national policy in respect of spent fuel and radioactive waste management;
   b. the significant milestones and clear timeframes for the achievement of those milestones in light of the over-arching objectives of the national programme;
   c. an inventory of all spent fuel and radioactive waste and estimates for future quantities, including those from decommissioning, clearly indicating the location and amount of the radioactive waste and spent fuel in accordance with appropriate classification of the radioactive waste;
   d. the concepts or plans and technical solutions for spent fuel and radioactive waste management from generation to disposal;
   e. the concepts or plans for the post-closure period of a disposal facility's lifetime, including the period during which appropriate controls are retained and the means to be employed to preserve knowledge of that facility in the longer term;
   f. the research, development and demonstration activities that are needed in order to implement solutions for the management of spent fuel and radioactive waste;
   g. the responsibility for the implementation of the national programme and the key performance indicators to monitor progress towards implementation;
   h. an assessment of the national programme costs and the underlying basis and hypotheses for that assessment, which must include a profile over time;
   i. the financing scheme(s) in force;
   j. a transparency policy or process as referred to in Article 10;
   k. if any, the agreement(s) concluded with a Member State or a third country on management of spent fuel or radioactive waste, including on the use of disposal facilities.
2. The national programme together with the national policy may be contained in a single document or in a number of documents.

Article 13 **Notification**

1. Member States shall notify to the Commission their national programmes and any subsequent significant changes.

2. Within 6 months of the date of notification, the Commission may request clarification and/or express its opinion on whether the content of the national programme is in accordance with Article 12.
3. Within 6 months of receiving the Commission's reaction Member States shall provide the requested clarification and/or inform the Commission of any revision of the national programmes.
4. The Commission, when deciding on the provision of Community financial or technical assistance for spent fuel and radioactive waste management facilities or activities, shall take into account the Member States' clarifications and progress regarding the national programmes.

Article 14 **Reporting**

1. Member States shall submit a report to the Commission on the implementation of this Directive for the first time by 23 August 2015, and every 3 years thereafter, taking advantage of the review and reporting under the Joint Convention.
2. On the basis of the Member States' reports, the Commission shall submit to the European Parliament and the Council the following:
   a. a report on progress made with the implementation of this Directive; and
   b. an inventory of radioactive waste and spent fuel present in the Community's territory and the future prospects.
3. Member States shall periodically, and at least every 10 years, arrange for self-assessments of their national framework, competent regulatory authority, national programme and its implementation, and invite international peer review of their national framework, competent regulatory authority and/or national programme with the aim of ensuring that high safety standards are achieved in the safe management of spent fuel and radioactive waste. The outcomes of any peer review shall be reported to the Commission and the other Member States, and may be made available to the public where there is no conflict with security and proprietary information.

Chapter 3 Final Provisions

Article 15 **Transposition**

1. Member States shall bring into force the laws, regulations and administrative provisions necessary to comply with this Directive before 23 August 2013. They shall forthwith inform the Commission thereof.

   When Member States adopt these measures, they shall contain a reference to this Directive or shall be accompanied by such reference on the occasion of their official publication. The methods of making such reference shall be laid down by Member States.

2. The obligations for transposition and implementation of provisions related to spent fuel of this Directive shall not apply to Cyprus, Denmark, Estonia, Ireland, Latvia, Luxembourg and Malta for as long as they decide not to develop any activity related to nuclear fuel.
3. Member States shall communicate to the Commission the text of the main provisions of national law which they adopt in the field covered by this Directive and of any subsequent amendments to those provisions.
4. Member States shall for the first time notify to the Commission the content of their national programme covering all the items provided for in Article 12 as soon as possible, but not later than 23 August 2015.

Article 16 **Entry into force**

This Directive shall enter into force on the 20th day following its publication in the *Official Journal of the European Union.*

Article 17 **Addressees**

This Directive is addressed to the Member States.

Done at Brussels, 19 July 2011.

For the Council

The President

M. SAWICKI